Applications of Geospatial Technology

Applications of Geospatial Technology

Edited by :

A. Ganesh

SATISH SERIAL PUBLISHING HOUSE
403, Express Tower, Commercial Complex, Azadpur, Delhi - 110033 (INDIA)
Phone : 011-27672852 Fax : 91-11-27672046
E-mail : info@satishserial.com, hkjain1975@yahoo.com
Website : www.satishserial.com

Published by :

SATISH SERIAL PUBLISHING HOUSE
403, Express Tower, Commercial Complex, Azadpur, Delhi - 110033 (INDIA)
Phone : 011-27672852 Fax : 91-11-27672046
E-mail : info@satishserial.com, hkjain1975@yahoo.com

ISBN 81-89304-25-9

Printed at
SALASAR IMAGING SYSTEMS, DELHI-35

Preface

The applications of geospatial technologies are playing an ever-increasing role in conservation and management of Advances in the science and technology behind remote sensing are making it more practical to conduct large-scale inventories and provide monitoring and evaluation of natural resources. Coupled with recent advances in GIS and spatial decision support systems, geospatial technology is taking resource conservation planning and management to a new level. Applied Geospatial Technology covers the fundamentals of techniques for acquiring, processing, classifying, visualizing and analyzing spatial-referenced data of the Earth, and their application to study of the Earth's surface features. Included are techniques such as land survey, photogrammetry, global positioning systems (GPS), airborne and satellite remote sensing, digital image processing, GIS and geovisualisation. In 2004 the US Department of Labor identified geotechnology as one of the three most important emerging and evolving fields, along with nanotechnology and biotechnology. Keeping this in mind, an attempt has been made in this publication for analyzing various facets of geospatial technology.

I am extremely thankful to Dr. C. Thangamuthu, Vice-Chancellor of the Bharathidasan University, Tiruchirappalli for his encouragement and support for publishing this book. The partial financial assistance provided by the Bhawathidasan University, Tinichirappalli for publishing this book is gratefully acknowledge. I extend my deepest appreciation to the contributors of this publication. I would like to thank Dr. K. Kumaraswamy and Dr. S. Aruchamy, Professors, Mr. R. Narayana Kumar and Mr. K. Banu Kumar, Guest Lecturers, Department & Geography, Bharathidasan University, Tiruchirappalli for their cooperation. I am gratefully acknowledged the technical assistance and support provided by Mr. R. Karvannan, Mr. R. Rajamani and Ms. S. Shanmuga Priya of the Department of Geography, Bharathidasan University, Tiruchirappalli. Finally my sincere thanks are also due to the Publisher, Satish Serial Publishing House for their excellent cooperation in bringing out this publication.

A. Ganesh

Contents

APPLICATIONS OF GEOSPATIAL TECHNOLOGY
A. Ganesh (Ed.), (2006) : pp.1-20
Satish Serial Publishing House, New Delhi
E-mail: ssp@ndf.vsnl.net.in, hkjain1975@yahoo.com
website: www.satishserial.com

1

Principles of GIS

P. ILANGOVAN

The Geographic Information System (GIS) has become ubiquitous today. Be it for the location of retail shops or for the operations of large corporate houses, GIS is invariably used to solve various spatial problems. Starbucks, a Coffee Shop Chain in London, uses GIS to locate its new branches and shortly Starbucks is coming to India. Reliance Telecom has used GIS to lay its cables and locate towers. Tata Indicom is using it for releasing telephone connection to a new customer. Indian Tobacco Company Ltd., (ITC) a multi-national company, has introduced a new scheme by name e-Choupal. By combining GIS and communication technologies, it is able to cover a number of states with its thousands of Kiosks to procure various agricultural products like wheat, soya bean etc for its factories directly from the farmers paying them a remunerative price. e-Choupal is shortly to be extended to all major states in the country covering more than one lakh villages and three million farmers to achieve a business transaction of over U.S.$ two billion. The archaeologists are locating new sites for excavations with the aid of GIS. All these have created a demand for GIS trained professionals. As a result, an increased number of institutions offer GIS courses of one kind or another. There is a growing awareness on the part of government organizations too about GIS. Local bodies apply GIS not only to enhance their revenue by efficient tax collection but also for their developmental planning. Indeed, many people do not recognize the term GIS, they are even less likely to be able to tell you how GIS has affected their everyday lives. But it has (Schuurman, 2004).

What is GIS?

GIS is a computer based information system to store, analyze and display of spatial data. Burrough (1986) defines GIS as 'a set of tools for collecting, storing, retrieving at will, transforming and displaying spatial data from the real world for a particular set of purposes'. A similar but slightly elaborate definition is given by the Department of Environment (1987) as 'a system for capturing, storing, checking, integrating, manipulating, analyzing and displaying data which are spatially referenced to the Earth'.

Historical Development

The roots of development of GIS go back to the 1960s. Parallel developments that took place in various fields led to the development of GIS. In 1962, McHarg a landscape architect with his 'overlay' technique paved the way for GIS. His technique has become the 'sine qua non' methodology for the present day GIS. This 'overlay' forms the basis for a range of analytical techniques broadly known as 'spatial analysis'. There was a growing interest in the late 1950s and early 1960s by the researchers in the U.S. to computerize mapping process. Influenced by the development of computers and quantitative revolution, tools began to be developed for spatial analysis. Tomlinson and Pratt were instrumental for the Canadian Geographical Information System (CGIS) in the year 1964. The CGIS was a pioneering work in which 'land use zones were digitally encoded so that they could be overlaid with other relevant layers' (Schuurman, 2004). Almost at the time, the traditional cartographers of U.K. and U.S. put up efforts to create automated maps. Quantitative Geographers like Berry, Tobler and Marble in U.S. and Waugh and Boyle in U.K. began to develop computer problems to solve spatial problems (Chrisman, 1988). During this time, Harvard Graphics Laboratory was established which paved way for effective vector analysis. Earlier in 1966, Harvard researchers had created SYMAP first raster GIS (Longley, 2000). A year later, U.S. Bureau of Census developed Dual Independent Map Encoding. Geographic Database Files (DIME-GDF). It was a database programme exclusively developed for 1970 Census to handle enormous amount of data.

In 1969, some of the researchers from Harvard started ESRI to undertake GIS Projects. A couple of years later, Landsat 1 was launched which paved the way for the relevant data of the resources of the earth's surface at a cost-effective price. In the year 1981, ESRI launched its first commercial GIS 'ArcInfo' which was designed for mini computers and based on relational database model. Since then much water has flown under the bridge. Computer hardware and software have become cheaper and more powerful. As a result, GIS became accessible to many. GIS industry has grown manifold in the recent years. Today's GIS business is worth more than U.S. $ 7 billion. It has more than a million core users and 5 million casual users. This impact is also felt in India which has more than one lakh GIS executives working in various business concerns. The GIS based BPOs pay more than voice based BPOs. High resolution satellite data, GPS Technology and developments in Information and Communication Technology have resulted in the generation and dissemination of high quality spatial data. The growth of Internet and the introduction of Internet Map Servers have resulted in the prolific use of GIS by people who know nothing about GIS. An average American Citizen before stepping out of his house for his work, checks Internet for the conditions of weather and the traffic. The data for these are constantly provided by the satellites and are processed by computers that are distributed over many places and feed the Internet on real time basis. WEB GIS and VIRTUAL GIS are the order of the day and the world is witnessing rapid changes in these hi-tech areas that will completely revolutionize the everyday life of the common man.

Components of GIS

Explaining the components of GIS is as difficult as defining GIS. Components that make GIS may be classified into so many ways. According to Burrough (1986) GIS components include computer hardware, a set of application software models and a proper organizational context. Heywood et. al. (2003) have included five components to describe the constituents of GIS: computer system, software, spatial data, data management and analysis, procedures and the people to operate GIS. Similarly, hardware, software, data, people and procedures and network constitute the components of GIS (Longley et. al., 2001).

Hardware components: Hardware components include all these gadgets and their accessories that sit on the desk, or on the lap or in the palm. Earlier, to carry out complex GIS operations, large mainframe computers were required. To day users have wider choices. Even moderately priced PCs are capable of being used for GIS. A range of latest products includes handheld GIS compatible GPS, a variety of high-end scanners and printers that make the GIS operations worthy of it. Now GIS functions are carried out in PDAs, in-vehicle devices and even in mobile phones.

Software: Software is required for various functions to be carried out in GIS. Software for data entry, data storage, data processing and analysis and data output are bundled together as software packages. These ranges from a simple package to work horses. Each product comes with different requirements and applications. Now freeware are available which can be downloaded from the Internet. The consumers can use these freeware not only free of cost but also change the source codes of the programmes. On the other hand, the ones available in the market are property right products whose source codes are protected by the Copyright Act and are not available for the users. Most of the software packages available commercially come as menu driven which are more user-friendly. Open GIS and Interoperatability are the recent keywords in the GIS world. Open GIS stands for open source coding. Interoperatability aims to increase interaction between different software packages for easy data transfer and processing from one package to another and vice versa.

Data: The third component of GIS is data. Data forms the major component. With recent developments in remote sensing and GPS technologies, large volume of high resolution data is widely available at affordable cost. This is also one of the reasons for the increased awareness and development of GIS. To store and maintain large volumes of data the use of database management systems is required. Till recently, Relational Data Base Management System (RDBMS) was widely used in GIS packages. RDBMS was not exclusively created to handle spatial data. As a result, Object Oriented Data Base is increasingly used in the recent years for its ability to handle spatial data more efficiently.

People and Procedures: A GIS is of no use if people and the organizations in which they work are not properly oriented towards GIS. In many organizations, GIS is not optimally used in spite of costly hardware and sophisticated software packages are available since the organizational aspects are not properly looked into. An organization must establish procedures, lines of reporting, control points and other mechanisms for ensuring that its GIS activities stay within budgets, maintain high quality and generally meet the needs of the organization (Longley et. al. 2001). The people working in an organization are also important. They possess various skills to handle geographic data. They design, programme and maintain GIS. They also supply data, do suitable analysis and interpret them.

Network: Earlier information generated in GIS is communicated among a handful of people sitting around a monitor. Today, a sheer volume of digital information shared among people who are located across the world. Internet plays a vital role to exchange and establish connection between them. Internet is not only used for personal exchange of information but also for the exchange of corporate data.

Representation of Geographical Features

For thousands of years, maps are used to represent the Earth. Maps are the most efficient and familiar way of representing complex earth on a two-dimensional surface namely map. The features of the earth's surface are represented either as objects or field points of view. In the objects point of view, different features grouped as points, lines and areas are represented as though they exist in an empty space. The features such as wells, trees, farmhouses etc. can be thought of as points even though they are three-dimensional objects on the surface of the earth. They are shown by symbols which are non-dimensional in character on the map which is two-dimensional in nature (Figure 1). Same way, roads or electrical lines or rivers are represented as a linear feature even though they cover an area. They are shown by uni-dimensional symbols on the map. Cropped area, forests, water bodies etc. are shown as area features. They are considered as two dimensional in character.

Field points of view conceptualize the earth's features as a continuous surface with infinitive number of points. Usually terrain is thought of as a continuously varying surface in which elevation can be defined at every point. The field view represents the real world as a finite number of variables, each one defined at every possible location (Longley et. al. 2000). Discrete objects and fields represent two conceptual views of spatial phenomena. But they are rarely useful in representing these features digitally in a computer which has its own limitations. Fields or objects have infinitive number of points or information whereas computers can take only finite number of points or information. Hence there is a need to device methods by which earth's features can be digitally represented in a computer. Raster and vector are the methods to represent the geographical features in digital computers. These methods reduce geographical features in such a way they can be made computer compatible. Practically raster method is strongly associated with field points of view and vector is highly associated with discrete objects.

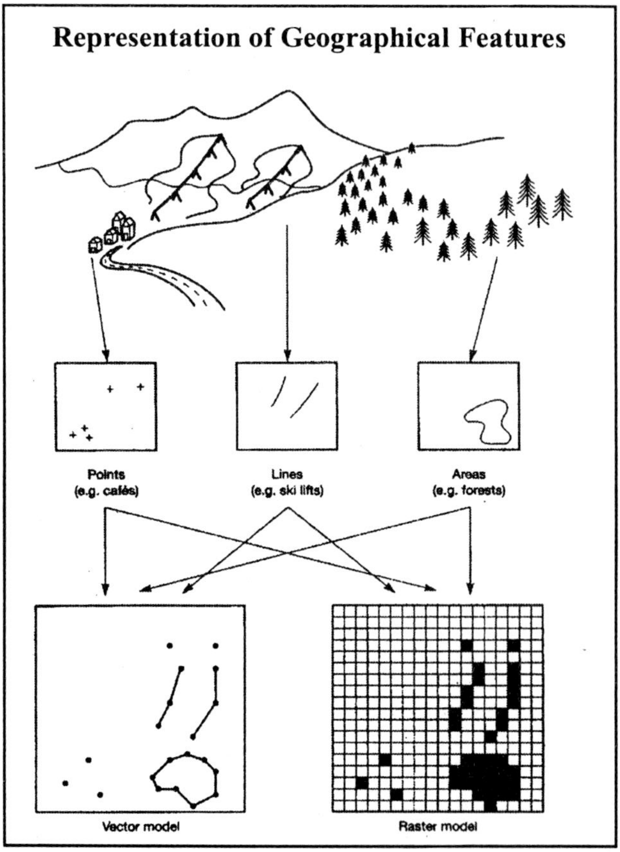

After Haywood et al., 2003

Figure 1

Raster Method

In raster, geographic space is represented by an array of cells or grids (Figure 2). The cells are the building blocks to represent various features on the earth's features. Raster model divides the world into array of cells and assigns attributes to these cells. In this method, a point is represented by a single cell; a line is by a series of cells in a given direction and an area by a group of cells. Because each cell can hold only one value, different attributes/ themes are shown in different data layers.

Raster method is easy to make and understand. The technology to make raster data is very cheap and hence vigorously developed. It was the earliest method to be developed because of its simplicity and cheapness. The raster data permits many arithmetical operations. For example, overlay analysis and simulation are very easy in the case of raster data. But the major problem with raster is its size. It is voluminous in nature. It requires large memory/storage space even for features occupying small area. As a result, it takes more time for the computer to process this voluminous data.

Raster and Vector Data Models

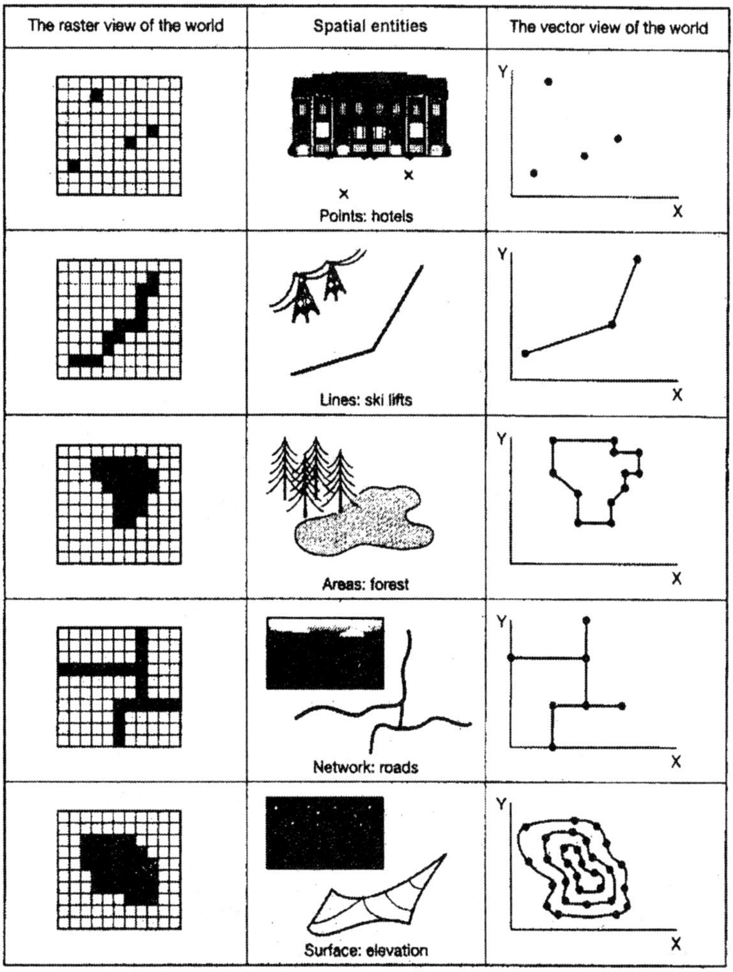

After Haywood et al., 2003

Figure 2

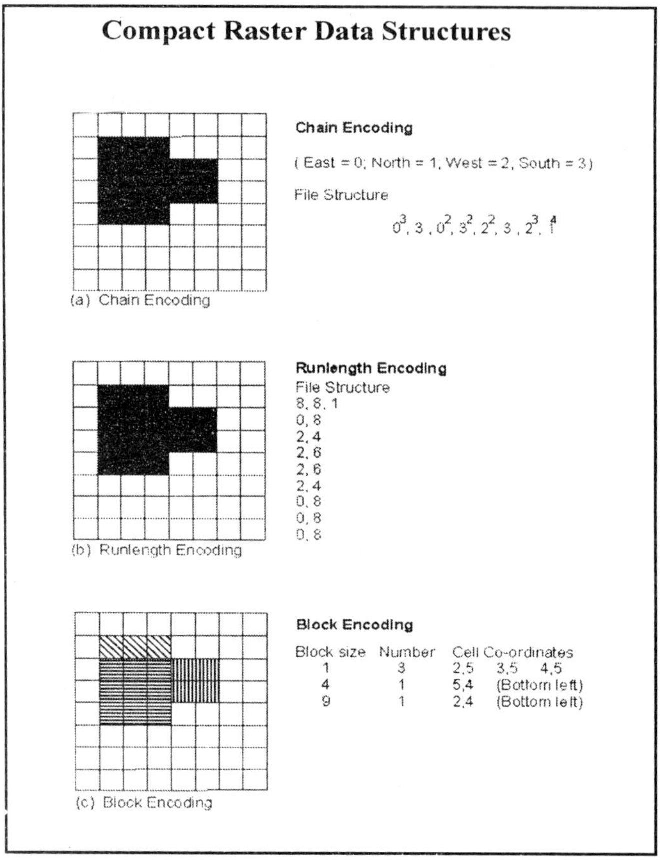

Figure 3

There are methods by which this voluminous data is made more compact so that it requires only small space in the memory. The following are some of the methods to compress the voluminous raster data to make it compact:

Chain codes

Run-length codes

Block codes and

Quadtrees

Chain codes permit to assess the area estimation and perimeters or detection of sharp turns and concavities made very easy. At the same time, redundancy is unavoidable in this method. Certain overlay operations like Union and Intersection cannot be performed without transforming chain codes into original data matrix.

Chain codes: In the case of chain codes, the boundary of the polygon/regions is represented in terms of its origin and sequence of unit vector in the cardinal directions. The numbers or alphabets may be used to indicate the directions. Number of steps/pixels in each direction is given by the superscripted numbers (Figure 3a).

Run-length codes: Run-length codes permit the cells of each row of the mapping unit to be stored from left to right with a begin cell and an end cell so that the empty cells may be deleted from storage (Figure 3b). By this method, only 19 cells are required to represent a feature that originally took 64 cells thereby effecting a saving of more than 70 per cent. This method is especially useful for small computers.

Block codes: In this method, the mapping unit is divided into different sized blocks. The data consists of three numbers, co-ordinate values and the radius of each square blocks (Figure 3c).

Quadtrees: One of the most compact ways of representing normal raster data is through quadtrees. It is obtained through subdividing each quadrant into further quadrants up to individual cells. Firstly the region is divided into four quadrants. In the given region (Figure 4), only first three large quadrants namely 0,1 and 2 have information. The fourth quadrant has no information and it is left out. The 0 quadrant is further subdivided into 00,01,02 and 03 quadrants. 03 quadrants have information in all cells. So no further subdivision is necessary. The other quadrants like 00 need further subdivision up to the cell level. The quadrant 00 is subdivided into 000, 001, 002 and 003. In this, the first three quadrants namely, 000, 001 and 002 fall outside the region and are represented by the empty cell ( ). Only 003 has information and hence it is represented by the pixel (ÿ). Likewise all the other quadrants are subdivided and represented by the quadtree method. Because of its efficiency, it is widely used in GIS.

Vector Method

In vector model, co-ordinate geometry is used to define the spatial entities. A point is represented by a single pair of x, y values. A straight line is represented by a starting x, y pair and an ending x, y pair. An arc or a chain is shown by means of a series of x, y pairs. A region or an area or a polygon by a series of x, y pairs where the starting and ending pairs are the same. When compared to raster data, vector data is more compact as it uses only a very little of computer's storage or memory. It also represents the features or objects as exactly as possible. Some of the disadvantages of the vector mode are that it has a complex data structure and many mathematical manipulations are not easy. Modification of the data is very difficult. Nevertheless, vector mode is widely used in GIS after the continued development of graphics in the computer field. In the recent years, vector analysis is highly developed.

Vector Data Files: Like raster data model, vector data also can be created and stored in various ways as simple data structures or topologically integrated data structures. Unlike raster data where attributes or themes are stored in different data layers, vector data model holds spatial data and attribute data often in separate tables.

Point Entities: Point entities are those, which are denoted by single pair of x, y co-ordinates. On the map they are represented by point symbols. Hence in addition to their x, y positions, the size and type of symbols and many other attribute information are also required to be stored in data files for display purposes.

Line Entities: In vector data files, line entities are stored as a pair of x, y co-ordinate values. As in the case of point entities, not only spatial data but also attribute data such as type of line symbol, name, colour etc. are also stored.

Area Entities: Different methods are available for storing area entities. Spatial and attribute data about area or polygon or region may be represented as simple polygon or with pointer or topologically integrated data structures.

The simplest vector data structure to show an image is a file containing the x, y co-ordinates of points of that image (Figure 5a). One of the problems of simple data structure containing individual polygons is the lacking of information about adjacency. If the co-ordinate pairs are created and stored separately, the lines common to some polygons have to be digitized and stored more than once. This creates room for errors as digitizing the same for more than once may not match with each other and will create problem when displayed. Also there is data redundancy when the same information is stored more than once.

In order to avoid some of the drawbacks of the simple polygon data structure, pointers are added to it. In this case, all co-ordinate pairs are sequentially numbered and are referenced in such a way that records which points are associated with each polygon (Figure

5b). It is an improvement over simple polygon data structure but many problems associated with simple polygon data structure still persist.

The aim of a polygon data structure is to be able to describe the topological properties of areas (shapes, neighbors and hierarchy) in such a way the associated properties of these basic spatial building blocks can be displayed and manipulated as thematic map data (Burrough, 1986). A topologically integrated polygon data structure satisfies this requirement. These data structures ensure no node or line segment is duplicated; line segments and nodes can be referenced to more than one polygon; all polygons have unique identifiers and island and hole polygons can be adequately represented.

Compact Raster Data Structures

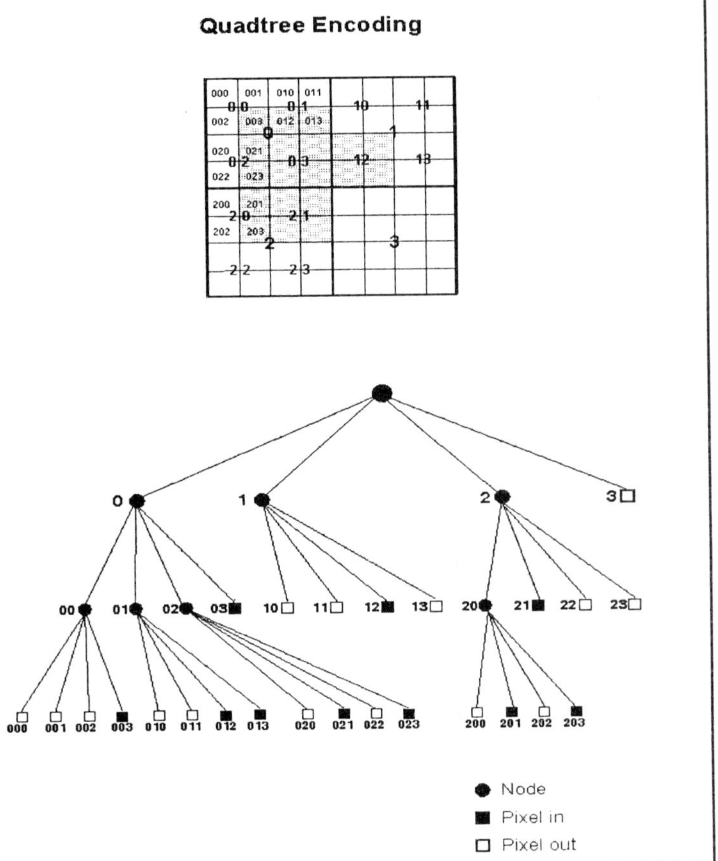

Figure 4

Raster or Vector?

A basic question arises often in the minds of students of GIS is which of these data models is the best? To find an answer for this question is very difficult. Both methods are complimentary to each other. Each method has its own merits and limitations. Earlier with limited capabilities, most of the hardware and software are designed to handle raster data only. Now-a-days with increased hardware capabilities and versatile software, the handling of either of the data is not at all a problem. Raster data and its analysis are simple and efficient for certain manipulations whereas vector data calls for greater sophistication for data analysis. Now vector data can be rasterized and analyzed and again vectorized. The same is applicable to raster data also. But conversion of one mode to another results in the loss of information and increased generalization. One should be aware of the limitations of these models and apply his mind in selecting the method and mode of data analysis depending upon the requirement of the problem.

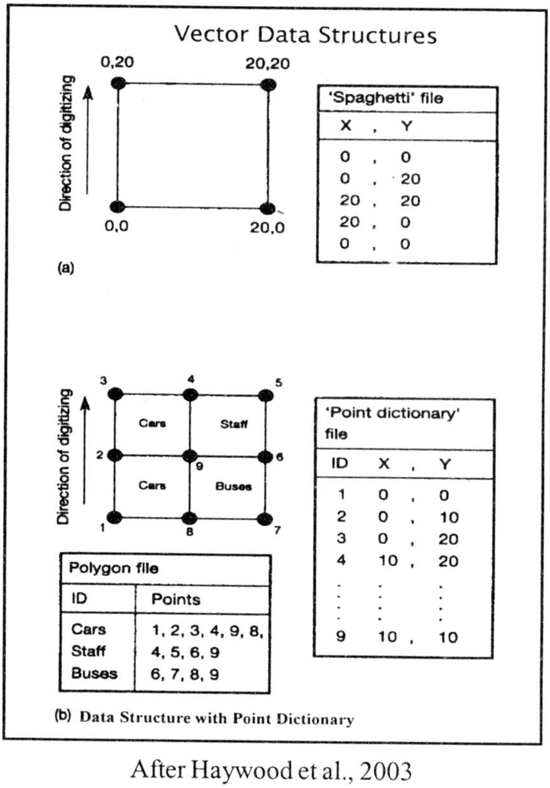

After Haywood et al., 2003

Figure 5

Data Quality and Storage

Data quality is important in GIS as it affects the quality of output. If the data used is of poor quality, the quality of results will also be questionable in nature. The errors in input data may be compounded during data processing and analysis. Hence one should be sure of the data quality before using the data in GIS so that results can be interpreted with confidence. In GIS data quality is used to give an indication as to how good the data are (Heywood et. al., 2003). Data quality implies the suitability of data for a particular purpose. It also indicates the extent to which data is free from errors and other problems. Errors, accuracy, precision and bias affect the data quality.

Errors result when data entered are inaccurate. Accuracy is the extent to which an estimated data value approaches its true value (Aronoff, 1989). It is impossible to have 100 per cent accurate data, but it is possible to have data that are accurate at acceptable levels. Precision is the recorded level of detail of your data (Heywood et. al., 2003). A level of precision does not mean a high level of accuracy. Human and technical sources of bias contribute to certain consistent errors in GIS. Human beings involved in interpretation consistently may omit features below a particular size. Often shortcomings in the software programmes and instruments not calibrated properly result in consistent bias. Bias can be very easily rectified but it is very difficult to locate them. In addition to errors, accuracy, precision and bias, data quality is also affected by inherent features of data sources like resolution and generalization (Heywood et. al., 2003).

Data Storage: Large scale GIS projects generate large volume of data that are required to be preserved in large data medium for many years to come. Archiving of mass of data is necessitated because of large scale digitizing activities carried out by National Mapping Agencies and other corporate houses. Magnetic tapes or disks are the usual medium of storage. Proper handling and storage of these storage devices ensure long life for the data stored. Dust free cool environment, storing them away from external magnetic and electrical sources and protection from fire are some of the essential requirements of storing data in magnetic tapes and disks. Now-a-days several new safe storage devices are available to store and preserve large volume of data. They are more compact and the data can be easily read and transferred to other devices.

Data Analysis in GIS

Data analysis forms the heart of GIS. The success of GIS lies in its ability to convert vast array of data into meaningful information. Most of the GIS packages can perform a wide variety of functions for the data analysis. These include measurement techniques, attribute queries, proximity analysis, overlay operations and the analysis of models of surfaces and networks (Heywood et. al., 2003).

Measurement techniques: Lengths of lines, distance between points, area, perimeter etc., are commonly measured in GIS. When a query such as what is the distance between A and B? or how far is B from A? measurement functions are used. The answer again would depend upon the type of GIS (vector or raster) and method of measurement used. The measurements made from GIS are always approximate values. In raster method, more than one answer is possible. The shortest path or 'Euclidean distance', 'Manhattan distance' and 'Proximity distance' may be derived which are different from each other. Perimeter is arrived at using raster method by multiplying the number of cells that constitute the boundary of an object or a feature and the cell resolution. Area is calculated by multiplying the area of the individual cell and the number of cells a particular feature occupies.

In a vector GIS, Pythagorean theorem is used to get the Euclidean distance. The perimeter is the sum of straight line segments.For calculating area, the feature is subdivided into simple geometrical shapes such as triangles and the area of each one is calculated and summed up. In Vector GIS, the length, area and perimeter are calculated and are stored as attribute data whenever a feature is created. They need to be calculated only once and they are stored permanently.

Queries: Queries are essential part of GIS. Role of queries is manifold. Query is a method of data retrieval. If the query is pertaining to a data that already forms part of the database, the data is simply retrieved and displayed. If the data is not readily available, new data is created by integrating relevant existing tables and the query is answered. Query is also useful as a method of checking the quality of data. With the help of a query, one can check whether the data is encoded and entered properly.

Queries are of two types, such as spatial and non-spatial queries. Spatial query is answered through analysis of spatial data. Non-spatial query pertains to attribute data analysis. The answer for such a query is made from attributes stored in two-dimensional tables. Boolean operators such as AND, NOT, OR and XOR are used to answer queries where queries are characterized by two or more spatial or non-spatial criteria.

Proximity or Neighborhood Analysis: A variety of functions are available to find out the influence of an entity/feature on its neighbors and the neighbor's influence on an entity. Buffering is often used to measure the extent of influence around an entity. Filtering is another function in proximity analysis.

Buffering is a highly useful technique that has many applications. Openshaw et. al.,(1989) have used buffering along with many other data layers to find out the sites for the safe disposal of radioactive wastes. One can create buffer zones around a point, line or an area. To understand the zone of interest around a point, a circle of required radius is drawn around that point. If the commercial banks serve 10 km distances around them on an average, then a circle is drawn around them according to the scale of the map for the distance. The circle indicates the service area or zone of influence of the bank. Likewise, circles may be drawn around banks in the locality. The area left out by the circles indicates the unserved area by the banks. New branches or banks may be located in these areas.

Buffering along a line can be created using the buffering function available in GIS. When one wants to know the number of houses to be demolished due the widening of the road by another 5 metres on either side GIS will create buffering zones or the required width on either side of the road and the number of houses that fall within the buffering zone can be calculated. In the same way, buffering zone around area features can be created.

Buffering indicates the influence of a feature on its neighboring areas whereas filtering assumes the influence of neighbors around an entity. Various types of filters and a number of algorithms are available to calculate the value of target cell/point in the case of raster data.

Map Overlay: The key analysis in GIS is the overlay analysis. In overlay analysis, a number of data layers are integrated to create a new data layer output. Raster overlay is conceptually simple and hence it is widely used. On the other hand, vector overlay is complex, time consuming and hence more expensive.

Vector Overlay: Vector overlay makes use of topology and geometry. Topologically correct data layers are needed for carrying out vector overlay analysis. When two data layers are overlaid, a new data layer is created. To measure the lengths of lines and areas, the geometry is used.

Point-in-polygon overlay is used to know the polygon in which a point falls. For e.g., when two data layers representing meteorological stations (point data) and land use (area data) are overlaid the resulting data layer will reveal which are meteorological stations that lie within the forest area and which fall outside the forest area (Figure 6).

Line-in-polygon overlay results when a query is raised to know which are the roads that fall within the forest area? Then the road data layer (line data) and forest polygon data layer are merged to form a new data layer. The new data layer provides the answer for the query raised.

Polygon-on-polygon Analysis: When the interest is to know the areas of forestry in the resort, the forest data layer and resort data layer are overlaid. Three different outputs based on Boolean Algebra or Mathematical Set Theory is possible based on the type of questions raised. If the interest is to know the areas of forest or areas of resort, then the answer will be based on Boolean OR operation or UNION (Set Theory). In the same way, if the query is to know the area of the forest within the resort, it results in NOT (Boolean) or ERASE (Set Theory) operation. AND (Boolean) or INTERSECT (Set Theory) operation will be performed when the interest is know the area that is both forest and resort.

Raster Overlay: Unlike vector overlay, raster overlay analysis is simple and straightforward. Many arithmetical operations such as addition, subtraction, multiplication and division may be performed between different layers to produce output (Figure 7). Standardization of data and weightings are also possible to make the analysis more meaningful (Heywood et. al., 2003).

Rasterization of different entities such as points, lines and areas will lead to the creation of different data layers. A layer with point data may be merged with area data that will result in a new data layer. If the interest is to know the locations where both meteorological stations in resort area are found, the analysis will reveal the cells that contain the information. Similarly a data layer containing line data may be integrated with that of the area data to create a new raster layer for answering certain queries to know about the location of roads in the forest. Likewise polygon on polygon analysis for raster data can be performed and the results of the new layer are interpreted.

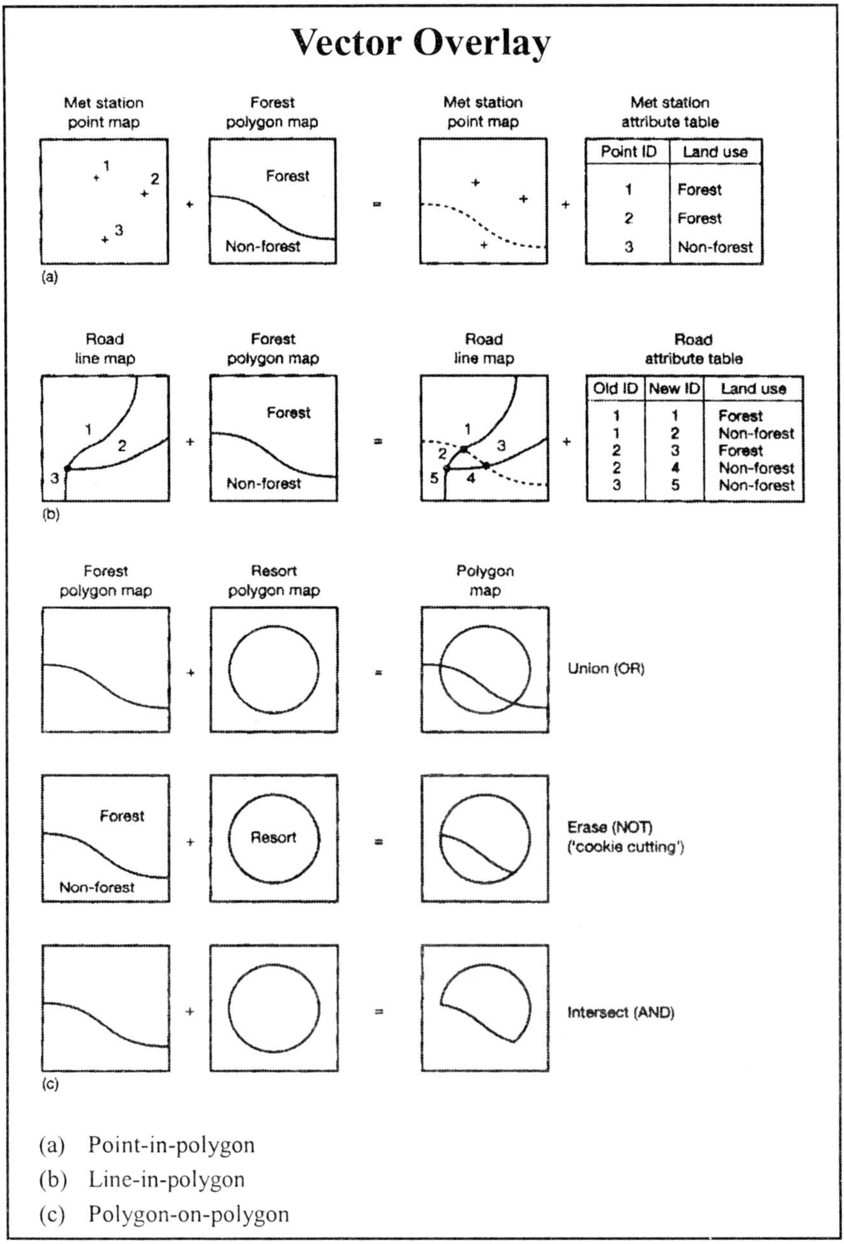

After Haywood et al., 2003

Figure 6

3D Modeling in GIS

One needs to visualize the shape of the features on the earth's surface for better understanding. Terrain is a three-dimensional surface which needs to be understood for many applications. For calculation of slope, aspect, visibility analysis and even for missile guidance systems three-dimensional view is essential. The digital representation of elevation/height is known as Digital Elevation Model.

Raster Overlay

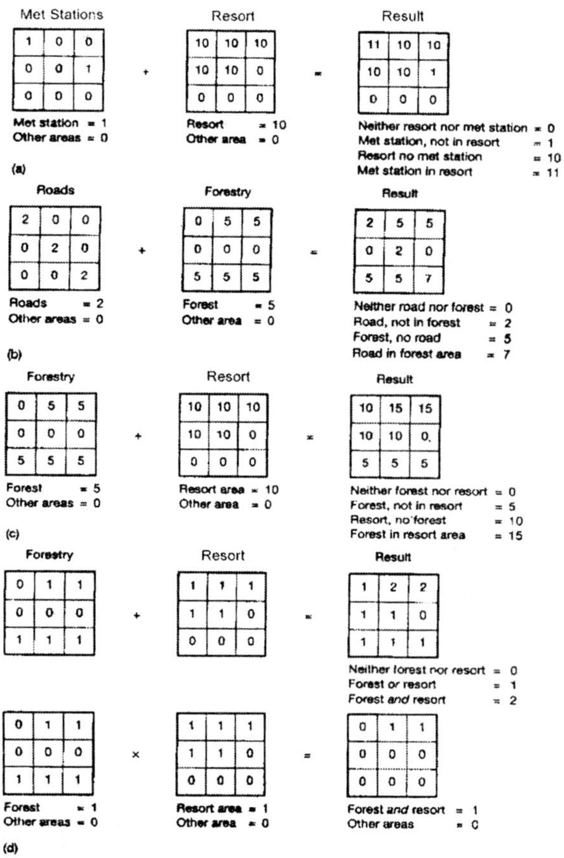

(a) Point-in-Polygon
(b) Line-in-Polygon
(c) Polygon-on-Polygon
(d) Polygon-on-Polygon (Boolean Algebra)

After Haywood et al., 2003

Figure 7

Digital Elevation Model. GIS can be used to create a DEM. Contours and Spot Heights are the common forms of the data for the construction of DEM. The maps, ground survey, aerial photographs, satellite images, field data including GPS etc. form the sources of data for the DEM. Both raster and vector methods can be used to create DEM. In the case of raster GIS, a DEM is derived from a grid of elevation values. For mountainous terrain, a finer grid of elevation values is required where more number of cells with values are needed to represent the varied topography more accurately. In plain areas, a coarse grid would suffice to create the surface. In areas where both rugged topography and plain areas are found, if one uses a finer grid, it will create redundancy in the plain areas i.e., same values in most of the cells. If one goes for a coarse grid to avoid data redundancy, mountainous areas will be poorly represented. This is a major drawback with raster GIS. Nevertheless, raster provides a simple method to create DEM.

The problem of data redundancy can be easily overcome by switching over to TIN (Triangulated Irregular Network) Model (Figure 8). The TIN model will have more elevation points from more triangles in areas of heterogeneous topography. In areas of homogeneous topography, a fewer number of triangles will sufficiently represent the terrain. In this way, varied terrain is more accurately represented. In the case of TIN, peaks and depressions are represented by vertices of triangles whereas ridges and valleys are represented by the sides (edges) of triangles. The surfaces of triangles provide area, gradient and aspect (Heywood et. al., 2003). All these data are stored in attribute tables and can be used for further analysis.

The terrain model thus created using raster or vector GIS have many applications. They are used to solve cut-in-fill problems for highway engineer who is interested to remove the ups and downs of the roads.DEM can be made more useful by draping any other information over it. The land use can be drapedover DEM to understand the land use better. Many other products like block diagrams, shaded relief maps can be derived from DEM. Location of hotels, resorts and watching towers both for civilian and defense purposes has been made easy through visibility analysis. Visibility analysis is done with the help of DEM. DEM is also highly useful for missile guidance systems. India's supremacy over its neighbors in the last Kargil War is largely attributed to the successful use of GIS in the war.

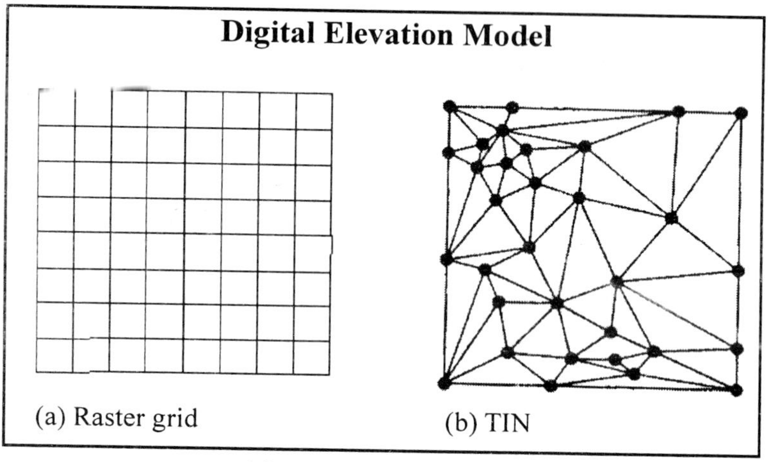

After Haywood et al., 2003

Figure 8

GIS has become a part of everyday life. It has both theoretical and practical applications. GIS has become a big business. Earnings from GIS software and industries providing GIS solutions have increased manifold. This indicates wider application of GIS in various fields and by many disciplines. A sound knowledge in various aspects of GIS and GIS packages will pave for better understanding of the world in which we live in.

REFERENCES

Aronoff, S., (1989). Geographic Information Systems: A Management Perspective. Ottawa: WDL Publications.

Burrough, P.A., (1986). Principles of Geographic Information Systems

For Land Resources Assessment. Oxford: Clarendon Press.

Chrisman, N., (1988). The Risks of Software Innovations: A Case Study of Harvard Lab, The American Cartographer, 15 (3), pp. 291-9.

Heywood, I., et. al. (2003). An Introduction to Geographical Information Systems. Delhi: Pearson Education Asia Pvt. Ltd.

Longley, P.A., et. al. (2001). Geographic Information System and Science. New York: John Wiley and Sons Ltd.

Openshaw S et. al. (1989). Britain's Nuclear Waste: Safety and Siting. London: Belhaven.

Shurrman, N., (2004). GIS: A Short Introduction. Malden: Black Publishing.

Tomlinson, R. F., (1988). The Impact of the Transition from Analogue to Digital Cartographic Representation, The American Cartographer, 15 (3): pp. 249-61.

APPLICATIONS OF GEOSPATIAL TECHNOLOGY
A. Ganesh (Ed.), (2006) : pp. 21-42
Satish Serial Publishing House, New Delhi
E-mail: ssp@ndf.vsnl.net.in, hkjain1975@yahoo.com
website: www.satishserial.com

2
GIS and Its Applications

S. VASUDEVAN

Right from the ancient days of mankind, representation of surface features was required as he moved about constantly in search of food, shelter and better hunting grounds or pastures. Surprisingly representing features as a map has been in practice from about 2300 B.C. Clay tablets excavated from Iraq show the outlines of a city map. Evidently, common public used maps as tools to represent and share information. Geographic surveyors, navigators and early explorers have made many efforts to collect map data for various purposes. Science of map making has undergone many changes and today it is a full-fledged discipline. During 1960's a new dimension of spatial data handling with respect to various resources and features came into existence. In the middle of 1960's, a few Geographic Information Systems (GIS) concepts were in vogue and familiar among the resource planners and evaluators.

Howard Fisher established a laboratory called Harvard Laboratory for Computer Graphics and Spatial Analysis at Harvard in the middle of 1960's. Initially, this lab developed a general purpose mapping software called SYMAP (SYnagraphic MAPping system) that had plenty of limitations when compared with today's software in all respects. Yet it was simple to use with a limited functional capability. It was very useful to non-cartographers to make maps. Subsequently the lab developed a few relatively advanced versions in the name of CALFORM, which was more cosmetic than SYMAP. SYMVU was another release in the late sixties by the same lab, which executed some Spatial Analysis with 3D perspective views over SYMAP. ODYSSEY was developed in the middle of 1970's, which was the first robust, efficient algorithm based package for polygon overlay.

Canadian Geographic Information System (CGIS) was initiated in the middle of 1960's to analyze the Canada Land Inventory and to generate statistics for use in planning land management activities for large areas of rural Canada. Its development contributed much to the conceptual and technical aspects of Geographic Information System (GIS).

Jack Dangermond founded Environmental Systems Research Institute (ESRI) in 1969. Work in ESRI was based on techniques and ideas developed at Harvard laboratory and elsewhere. One serious difficulty in earlier GIS software was that they could not handle attribute data. At the same time, during 1970's raster and vector based GIS models to various applications were made better and powerful. Reflecting this context, ESRI released a GIS software called ARC/INFO (ARC - Lines / Map features; INFO - Information about the line / map features). It was the successful marriage of standard database management system to handle both attribute tables with specialized software and objects stored as arcs. ARC/INFO was the first GIS software to take advantage of super-mini computer-hardware.

GIS stands for Geographic Information System. It is a particular form of information system where system refers to a group of connected entities and activities that interact for a common purpose. An information system is a set of processes, executed on raw data, to produce information (new), which will be useful in decision-making. Therefore, any information system must have a full range of functions to achieve its purpose, including observation, measurement, description, explanation, forecasts, etc. A Geographic Information System (GIS) uses non-spatial data and includes operations, which support spatial analysis.

In this context, GIS can be seen as a system of hardware, software, expertise, data and procedures designed to support the capture, management, manipulation, analysis, modeling and display of spatially referenced data for solving complex planning and management problems. It is better to further differentiate that even though other computer program can use spatial data (e.g. AutoCAD or any statistical package), GIS includes the additional ability to not only handle attribute data but also perform spatial operations.

Contributing Disciplines and Technologies

GIS is often termed as an "Enabling Technology" because of the potential it offers for the wide range of constituent disciplines. The following are some examples to substantiate it.

Geography

- Broadly concerned with understanding the world and man's place in it
- Long tradition in spatial analysis
- Provides techniques for conducting spatial analysis and a spatial perspective on research

Cartography

- Concerned with the display of spatial information
- Currently the main source of input data for GIS is maps
- Provides long tradition in the design of maps which is an important form of output from GIS
- Computer cartography (also called digital cartography, automated cartography) provides methods for digital representation and manipulation of cartographic features and methods of visualization

Remote Sensing

- Images from space and the air are major source of geographical data
- Remote sensing includes techniques for data acquisition and processing anywhere on the globe at low cost, consistent update potential
- Many image analysis systems contain sophisticated analytical functions
- Interpreted data from a remote sensing system can be merged with other data layers in a GIS

Photogrammetry

- Using aerial photographs and techniques for making accurate measurements from them, photogrammetry is the source of most data on topography (ground surface elevations) used for input to GIS

Surveying

- Provides high quality data on positions of land boundaries, buildings, etc.

Geodesy

- Source of high accuracy positional control for GIS

Statistics

- Many models built using GIS are statistical in nature, many statistical techniques used for analysis
- Statistics is important in understanding issues of error and uncertainty in GIS data

Operations Research

- Many applications of GIS require use of optimizing techniques for decision-making

Computer Science

- Computer-aided design (CAD) provides software, techniques for data input, display and visualization, representation, particularly in three dimensions
- Advances in computer graphics provide hardware, software for handling and displaying graphic objects, techniques of visualization
- Database management systems (DBMS) contribute methods for representing data in digital form, procedures for system design and handling large volumes of data, particularly access and update
- Artificial intelligence (AI) uses the computer to make choices based on available data in a way that is seen to emulate human intelligence and decision-making - computer can act as an expert in such functions as designing maps, generalizing map features. Although GIS has yet to take full advantage of AI, AI already provides methods and techniques for system design

Mathematics

- Several branches of mathematics, especially geometry and graph theory, are used in GIS system design and analysis of spatial data

Civil Engineering

- GIS has many applications in transportation, urban engineering, etc.

Data Types

Much of the data for a GIS pertains to spatial dimensions - having a physical dimension, geographic locations in terms of spatial coordinates on the surface of the earth and attribute data. A GIS essentially integrates the spatial and attribute data for planning, resource management and decision-making purposes, through well-defined operations. Spatial data have spatial dimensions and can be represented by a set of co-ordinates or geographic locations on the surface of the Earth. A map is an example of spatial data representing geographic features of the Earth. Non-spatial or Attribute data is data that qualifies the spatial data. It is quantitative and expressed as names, numbers, data values, etc. Census data on population, services, etc. are examples of non-spatial data. The data for a GIS could be categorized as follows (Rao, 1993):

a) Dichotomous or presence / absence: This type of data mainly is a logical definition of data characteristics and is also referred to as YES/NO data. It mainly applies where data is classified into one of two categories - a village may or may not have a school facility, a road may or may not have lanes, a city may or may not have an airport.

b) Categorical: This measurement scale is used when data is classified into one of several categories–different rock types, different vegetation covers and administrative categories etc.

c) Ranked: There are two types of ranked data. Ranked (or ordered) categories are used where a set of categories has a natural ranking associated with it - e.g. grades of agricultural land; alternatively a set of places may be ranked from first to last according to some criterion.

d) Count: Data consisting of a number of items or a number of times some thing has happened in a place – population, the number of species, the number of schools, the traffic count or the number of customers etc.

e)Co ntinuous: A measurement on a continuous scale, such as paddy production, average annual rainfall, height above sea level or the unemployment rate. Sometimes this may be reducible to the ratio or the sum of count variables, sometimes not.

Data Model

As the geographical variation in the real world is infinitely complex, it would take an infinitely large database to capture this real world precisely. However, data must somehow be reduced through a process of generalization and abstraction and at the same time, geographical variation must be represented in terms of discrete elements and objects. The data model is used to convert real geographical variation into discrete objects. Tsichritzis and Lachovsky (1977) define data model as 'a set of guidelines for the representation of the logical organization of the data in a data base'.

Present GIS differs according to the way they organize reality (real world) through the data model. They are Raster Data Model or Cartographic Map Model and Vector Data Model or Geo-Relational Model. Each model tends to fit certain types of data and applications better than other did. Most of the packages have the capability of handling both the data models.

Raster Model

Raster Model divides the entire area in to a regular grid in a specific sequence. It is generally sequenced row by row from top left corner. Each grid cell contains a single value. In most cases, the values are to be assigned to each and every grid in the raster data model. It is then written into a file, often coded in ASCII. It is relatively a simple approach to data integration, both conceptually and operationally and it remains popular since the earliest days of GIS development due to mainly its organizational simplicity.

The type of values contained in a raster, depend upon both the coded reality and the GIS. Different systems allow different classes of values including, whole numbers, real numbers, alphabetic values etc. Even though raster GIS allows several types of values, some layers numeric and some non-numerical, precautionary measures must be taken to avoid any unreasonable mathematical operations.

Resolution, the minimum linear dimension of the smallest unit of geographic space for which data are recorded is very important in terms of accuracy of the data. In the raster data model, resolution is limited to its cell size often decided by the user depends upon the size of the study area, purpose of the analysis, etc. High resolution refers to raster with small cell dimensions and vice versa.

The advantage over the data availability in this model is digital form of raster data, as images, pictures, etc. However, re-sampling will likely be needed in order to match pixel with other layers. Because remote sensing generates images, it is easier to interface with a raster GIS than a vector model.

Analysis in Raster GIS

Data Display

Displaying layers in a sensible manner is important for display. On a color display, each integer value can be assigned a unique color to differentiate the other value. When the values have some natural arrangements, the sequence of color is very important. For example, elevation is often shown on a map using the sequence blue-green-yellow-brown-white for increasing elevation. There should also be cartographic map elements like legend and north to fulfill the meaning of the map. Other type of display offers still clear understanding about the layer. This includes a perspective view by drawing profiles across the raster.

Local Operations

Local operations are nothing but value of each new pixel is assigned by the values of the same pixel of the input layer. In this operation, neighboring or distant pixels will not be changed or affected.

For example, assigning new values by assigning pixel to classes or ranges based on their old values is illustrated below.

0 - 499 - 1

500 - 999 - 2

> 999 - 3

New value can also be assigned as

New value = (2 * old value)

Operations on Local Neighborhoods

The value of a pixel on the new layer is determined by the local neighborhood of the pixel on the old layer. A filter operated by moving a window across the entire raster will produce two major effects. They are smoothing and edge enhancement. Smoothing which is called Low Pass Filter, removes finer / local detail while edge enhancement which is called High Pass filter, exaggerates finer / local detail.

Slopes and Aspects

If the pixel values in a layer are elevations, one can compute the steepness of slopes by looking at the difference between value of the pixel and of its adjacent neighbors. The direction of the steepest slope or the direction in which the surface is locally facing is called aspect. This can also be computed in degrees from north. This will be highly useful in analyzing vegetation patterns, computing energy balances and modeling of erosion and run-off. As aspect determines the direction of run-off, this can be used to sketch drainage paths for run-off.

Operations on Extended Neighborhood

Distance measurement of each cell from a particular cell or the nearest of several cells is possible. This is used for buffer generation around an object and features and is a very useful capability. For example, water logging buffer can be generated at 500 meters width around all lakes and water courses of a city. This will be useful for planning the buffer zone. Similarly, noise buffers around roads, safety buffers around hazardous facilities can also be created.

Operations on Zones

By comparing adjacent pixels, all patches or zones having the same value or range of values can be identified. Instead of using original pixel value, zone values will be used to represent the group of pixels.

Vector Model

In Vector data model, attribute information is always associated with point, line and polygons as spatial entities that describe features occurring in the real world. As spatial entities, they describe features occurring in the real world. Thus, for example, a point located feature such as a city may have associated with it such items of information as its total population, number of houses, schools and so on. Similarly, a linear feature such as a river might have associated with it such information as name, mean discharge, etc. A polygonal feature such as a landuse category might be linked to information describing its use, past land use, its soil type and so on.

The vector data structure represents each object by an array of coordinates. There are several variations possible. In a whole polygon structure, each data plane is sub-divided into further polygons. This method of storage is not optimal, as cognizance is not taken of shared lines and points. Modern vector based structures utilize this feature by building up a topology which relates the features by building from the arc-node structure consists of points or nodes. Arcs are lines joining nodes and polygons are areas completely enclosed by arcs. Data redundancy is reduced because shared nodes and arcs are stored only once. Attributes are linked to each feature. A refinement to this is the relational arc-node structure where the attribute data is stored in separate relational tables. The overhead here is more pointers and files to maintain but general purpose DBMS can be used for this purpose. In this case, access is more efficient.

Database Creation

Database creation in vector model involves three major stages. They are

First stage	:	Spatial Data Input
Second stage	:	Attribute Data Input
Third stage	:	Linking Spatial and Attribute Data

Spatial data or any thematic layer can be captured through digitized points and lines, scanned and vectorised points / lines or directly from other digital sources. Once the spatial data has been entered, much editing is still needed before it can be taken to further processing. Topology generation, a process by which problems such as overshoots, undershoots and spikes on lines are either flagged for editing by the user one by one or corrected automatically by accepting some tolerance values given by the user. Tolerance value is closely related to the precision of the data and therefore, these edit procedures include such functions as snap, move, delete, split, join etc. It is worthwhile to note that poor digitizing leads much need for editing and good digitizing can avoid most need for editing but both are labor-intensive works. Vectorisation over the scanned image has become more automated in joining lines and usually errors at later stages are less.

As a next stage, attribute data can be keyed in or imported from other digital database. Spatial entities are usually linked with their associated attribute data by means of a common spatial key called unique identifier (ID). Unlike in many raster systems, attribute data is stored and manipulated in entirely separate ways from the locational data.

Analysis in Vector GIS

Simple Display and Query

Point features can be displayed in the form of different markers with different color. Markers are taken out from a symbol library and values for markers either will be taken from attribute data file or will be keyed in by the user. Similarly line and polygon features are represented in the form of line patterns and polygon shades respectively.

Many a vector GIS uses Standard Query Language (SQL) which has standard phrase structure. SQL accommodates relational, arithmetic and Boolean operations to build a condition. This will be applicable to all point, line and polygon features.

Reclassify, Dissolve and Merge

These operations are used frequently on polygon features. These are used to aggregate area features based on attribute database. For example, one may wish to produce a map of major soil types from a layer that has polygons based on much more finely defined classification scheme.

Topological Overlay

Overlaying or combining features from two themes can create new spatial features and the technique is termed spatial overlay. Apart from the combination of the spatial geometry features (point, line, polygon), the attributes of the features are also combined to describe the attributes of the new theme. Three types of overlay operations are generally performed:

- Polygon overlay, where the polygons of two themes are merged or combined to create a new theme.

- Line-Polygon overlay, where the intent is to find out the polygons (of one input theme) in which line features (of second input theme) passes.
- Point-Polygon overlay is the process of deriving the information on polygons of one theme within which point features of another theme falls.

Spurious Polygons

During polygon overlay, many new and smaller polygons are created, some of which may not represent the spatial variations. These polygons are called spurious polygons or sliver polygons. It is generally a major problem in the polygon overlay.

Buffering

Buffering is an important function used to determine spatial proximity or nearness of various features by defining a distance zone around map features. Buffer can be generated for points, lines and polygons. Inward and outward buffers can be generated in case of polygons.

Feature Extraction

Extracting spatial features of a theme based upon the spatial geometry of another type is a function of a GIS. Say, if river system of a state is a theme and if the rivers of a district are to be obtained, then clipping the river features based on the district boundary could do this.

Feature Reduction

Slivers are small polygons that are resulted when an overlay operation is performed. These have to be removed before a proper analysis is done. Further, these small polygons would become "noise" in the spatial data and have to be merged with the neighboring polygon. This is dove as a feature reduction process by removing the largest shared border of the sliver polygon.

Transformation

Theme transformation pertains to the transformation of the co-ordinate system of a theme. The transformation is achieved through a minimum set of registration points specified in both the themes.

Raster versus Vector Data Model

There were many arguments and debates about which was better data model since the earliest systems were created. Raster databases are appealing because of the organization simplicity of spatial data and speed of many spatial operations like overlay, buffer, etc. It is especially very appealing to the remote sensing community who are used to pixel-based data processing. However, there are many situations in which the raster approach may appear to sacrifice too much detail. Cartographers ended up with the effect of diagonal boundaries represented by grid cell edges diagram. Surveyors were dismayed by the inaccuracy caused by the cells when portraying linear features and point features.

Coming to the Vector GIS, it is precise and has no approximation errors for measured quantities like area, length, perimeter, etc. It has less storage requirements. However computing times for overlaying vector-based information is excessive.

Today, there are situations in which it is clear that no approach is more functional than the other. For example, viewshed algorithms to find area visible from a point are feasible only with elevation grids and not with digitized contours. At the same time, land survey data can only be represented with precise lines and not by raster GIS.

An important current trend involves linking raster and vector systems displaying one over the other. Therefore, we have flexibility to use the most appropriate approaches on a case-to-case basis.

In this context, the following are to be considered before selecting a data model.

1. Coordinate Precision
2. Speed of Processing
3. Mass Storage Requirements
4. Characteristics of Phenomena

Coordinate Precision

In a raster model, locational precision is limited by cell size. Smaller the cell size, greater is the locational accuracy. However, location of a co-ordinate in any cell may be the center of the cell or one of its corners and the precise location of co-ordinate is ambiguous.

In Vector GIS, it can be encoded with any conceivable degree of precision. It is limited by the method of internal representation of coordinates (i.e.) single (8 decimal digits) or double (16 decimal digits) precision.

As for as data precision is concerned, vector precision is true for certain classes of data. For example, land surveyors coordinate, political boundaries are defined by accurate survey. A few natural phenomena have true edges, which can be accurately represented as mathematical lines like soil type, vegetation type, drainage line, etc.

Speed of Processing

Raster data can be processed very quickly to answer most analytical questions involving overlays, proximity and boolean queries. In most cases, analysis requires cell-by-cell comparison of the contents of layers. The raster systems are suitable for low program development costs. The same questions require considerable computation in vector systems because vector topology is more complex and complex geometrical problems must be solved, etc.

Mass Storage

To put it simply, raster storage method requires more space and vector need very little storage space. In raster, compression is possible in several methods but there is a small overhead in packing and unpacking data compared to cell-by-cell storage.

Characteristics of Phenomena

As raster is a regularly spaced sampling of phenomena, it reflects lack of knowledge of spatial variation. It is more appropriate for remote sensing data analysis. Vector representation permits more spatial variability. This is appropriate for social, economic, demographic, resources variation analysis, etc. To conclude this debate, one can store data in one form and process it in another. Data can be captured and stored in vector mod and analyzed in raster mode. This will considerably save mass storage space and computing time. The other way is using combinations of GIS system, which runs raster and vector analytical systems in parallel. For example, vector based landuse map can be overlaid on satellite imagery to improve the interpretation.

GIS Data Bases

As the real world is too complex for immediate and direct understanding, we create models of reality that are intended to have some similarity with selected aspects of the real world. Databases are created as a fundamental step from these models, of reality. Here, the database - spatial database is a collection of spatially referenced data-acts as a model of reality.

The GIS has two distinct utilization capabilities. The first is pertaining to querying and obtaining information and the second pertaining to integrated analytical modeling (Burroughs, 1988). However, both these capabilities depend upon the core of the GIS database that has been organized. Many a GIS utilization has been limited because of improper database organization. The importance of the GIS database stems from the fact that the data elements of the database are closely interrelated and thus need to be structured for easy integration and retrieval. The GIS database should also cater to the different application needs. Before the database organization, two major aspects are to be considered:

a) Organizational Commitment

An important aspect of any organization setting up GIS-based information systems is the level of commitment of the organization. This involves a decision at the organizational level committing itself to the GIS and defining the GIS as an amalgamated system of the organization and its work-culture. This would require an orientation/ training of senior decision-makers with specific emphasis on GIS and its capabilities; difference between GIS and other packages being used in the organization; what can be achieved by the organization using GIS; implications of GIS to organization's goals and objectives; assessment of effort; broad system definition and cost/ time/ manpower implications all should be subjected to a conscious scrutiny.

b) GIS Selection and Procurement

The process of evaluating a Geographical Information System (GIS) package is one of the prime activities for any organization or agency interested in GIS procurement. The evaluation process is a difficult one because of the complexity of the GIS technology and because of a variety of techniques and methods adopted for realizing the various functions of the packages.

Moreover, comparison of the techniques and methods used for the functions is not possible because of the non-availability of this information from the vendors. Further, there are a large number of GIS packages available in the market - making it almost impossible to perform a systematic evaluation.

Generally, a proper database organization needs to ensure the following:

a) Flexibility in the design to adapt to the needs of different users.
b) Controlled and standardized approach to data input and updation.
c) System of validation checks to maintain the integrity and consistency of the data elements.
d) Level of security for minimizing damage to the data.
e) Minimizing redundancy in data storage.

While the above is a general consideration for database organization, in a GIS domain, the considerations are more pertinent because of the varied types and nature of data that need to be organized and stored. Just as in any normal database activity, the GIS database also needs to be properly designed to cater to the needs of the application that proposes to utilize it. The design would also define a comprehensive framework of the database and allow the database to be viewed completely so that interaction and linkages between elements can be pre-defined and evaluated. The design would also permit identification of potential bottlenecks and problem areas so that design alternatives can be arrived at. The design would also help in identification of the essential and correct data elements to be incorporated into the database and filter out irrelevant data. Definition of updation procedures could also be a part of the design activity so that newer data can be incorporated in future and the database is active.

The design of the GIS database will include three major elements:

a) Conceptual design, basically laying down the application requirements and specifying the end-utilization of the database. The conceptual design is independent of hardware and software and could be a wish list of utilization goals.
b) Logical design, which is the specification of the database vis-à-vis a particular GIS package. This design sets out the logical

structure of the database elements and is determined by the GIS package.

c) Physical design, which pertains to the hardware and software characteristics and requires consideration of file structure, memory and disk space, access etc.

Each stage is inter-related to the next stage of the design and impacts the organization in a major way. For example, if the concepts are clearly defined, the logical design is easier done and if the logical design is clear, the physical design is also easy. The success or failure of a GIS project is determined by the strength of the design and a deal of time must be allocated to design activity.

Modeling in GIS

All problems of GIS based integration involve a conjunctive analysis of multi-parameter data. The multi-parameter data include different spatial inputs - maps of landuse, soils, slopes, terrain etc. - and other non-spatial datasets. The GIS allows for the integration of these datasets to obtain a composite information-set. However, the important aspect is the interpretation and analysis of the integrated information-set. Towards this, two aspects are of importance:

a) A model/criterion that defines the logic for the analysis of the composite information-set. The criterion is basically a logical objective. The model may be as simple as reclassification of the parameter values (say merging different landuse/cover categories) or a more complex multi-parameter criterion.

b) The relative importance or weightage of each parameter for the end-objective.

Both these two aspects form the fundamentals of data integration.

Criterion for Analysis

A criterion is a definition of a relationship amongst the different parameters in commensurate terms i.e. expressing the different parameters in the same units or scale. This is an important aspect of the modeling as most of the GIS datasets are represented in different scales and relationships are not well defined. For example, if a soil map and a landuse map has been integrated, then the relationship of the different landuse and soil categories need to be defined. The criterion is also dependent on the aim or objective of the analysis. There could as many criterions as objectives but there is a need to be able to select the optimal solution. For example, the integration of landuse and soil information could be modeled to identify suitability for cash crops and suitability for plantations. However, decision on the actual use would be based on the optimality of the solutions. The relationship definition in all these multi-parameter integration is highly subjective and consensus rarely exists concerning the relationship. Thus, the approach needs to be to arrived at a consensus or adopt multiple criterions and select an optimal solution. A criterion could be categorized as follows:

a) Statement Criterion: Where the relationship of the parameters is defined as a statement and is a mundane definition of the criterion. An example,

IF	SOIL	=	deep. Or. Very deep
	LANDUSE	=	wasteland
	GROUNDWATER	=	good
THEN			
	SOLUTION	=	grassland or afforestation

In this method, the criterion is class dependent and the relationship of other classes are not defined. As a result, the criterion is too rigid. For example, the above statement does not define what happens to shallow soils, forest /agriculture landuse and poor groundwater categories. However, this method is appropriate where there is no ambiguity in the relationship.

b) Table Criterion: which is a more organized and systematic definition of the relationship amongst all parameters and its categories. The table criterion lays down all possible combination relationships and thus provides a total solution from the dataset.

Determining Weightages

One of the classic problems in decision theory or multi-parameter analysis (like the one for integrated GIS analysis) is the determination of the relative importance of each parameter. This is a problem, which requires human judgment supplemented by mathematical tools. For example, in a criterion, definition, the relative importance of each parameter is undefined and thus an implicit equality of parameter importance is assumed. This may not be the real case and the parameter importance may need to be defined to represent a more optimal model of the problem. The objective is usually represented by a set of weights, in other words preference information, which is translated into numbers and is normalized to a constant or unity.

A number of methods are available to deal with such problems. The choice of the weightage determination technique could be commensurate to the objective and the dataset. However, weightages for analysis could be as follows:

a) Equal weighatges, where parameters are treated to be equal in importance and thus characterization becomes difficult and analysis solutions are not gradable.

b) Unequal weightages, where the importance of some parameters over others is characterized and the solution is gradable and categories could be defined.

c)S ingle parameter weightages, where a single parameter is weighed to unity and all other parameters are treated as 0. Thus, the solution is totally influenced by the single weighted parameter. This case is adopted to characterize the worst or best parameter in a dataset and to define a solution around this parameter.

Digital Elevation Models

Digital Elevation Model (DEM) is frequently used to refer to any digital representation of a topographic surface. Digital Terrain Model (DTM) is more generic term but it is not widely used. The resolution of a DEM - the distance between adjacent grid points - is a critical parameter in DEM. Different methods have been used to create DEM. DEM can be generated from contour lines of a terrain. Contours are vectorised and they are tagged with their respective elevation values. Finally, an algorithm is used to interpolate elevations at every grid point from the contour data. DEM can also be generated using photogrammetric techniques. Parts of stereophotos through a stereo plotter have been used to bring same positions until they appear to be one. When processed completely, the instrument calculates the parallax displacement of a larger number of points. In this way, elevation data can be acquired from photographs. DEM is useful in determining attributes of terrain such as elevation at any point, slope and aspect. Features that are lying on the terrain such as drainage basins, watersheds, viewsheds, drainage network, channels, peaks, pits and other landforms can be demarcated through DEM. Further hydrological modeling can be performed using DEM data.

Applications of GIS

Various parameters viz. structural, geomorphological, subsurface geological and aquifer characteristics parameters have been used to develop a aquifer function modeling in Vellar basin of central Tamil Nadu. Map data have been digitized and GRAM (Geo Referenced Area management)- a raster based GIS was used for GIS modeling work. The results of the analysis show that several different zones have got varying behavior among all the parameters used for this study. In this study, most of the operations are confined to overlaying thematic information one over the other. Reclassification procedures were frequently used before proceeding to next step of analysis (Vasudevan, 1995).

Mumbai Metropolitan Region development Authority (MMRDA) has made urban development study by using large scale map. Most of the large scale maps (1:8,000, 1:5000) have been collected from the office of Land Records, Greater Mumbai. In this urban development study, socio economic facilities, agriculture, forest landuses, soil type, altitude data have been enormously used. Land suitability analysis carried out under this program shows landuse zones marked for residential, commercial, industrial, openspaces, etc. It also indicates the existing road, water bodies, propose roads for future expansion (Patkar and Kumar, 1999).

Routing analysis for the proposed airport in BMR based on an evaluation of travel-distances/shortest-path and the travel-times on the existing road network and a network including proposed freeways. Use of NETWORK analysis in urban planning is important for traffic and route analysis. This also has an impact on siting of urban facilities. As a demonstration of the route analysis, the NETWORK module has been used for determining routes from various locations in BMR to the proposed airport near Mandwa. The route analysis determines the shortest distance and travel-time on the existing and proposed road network of BMR (Anon., 1992).

Wasteland Development to suggest a plan for the development of the wastelands has been done for Aspur tehsil, Dungarpur. Towards this, spatial patterns have been generated to serve as indicators and provide inputs towards the planning process. Integrated analysis for wasteland development planning is based on the integration of the people's needs, identified and preferred use to which the land could be put to. Based upon this the implementation plan is generated. Panchayat based plan input generation, which includes the development measures and the possible use to which the land unit can be put to.

Decision Support System at taluk level was attempted as a pilot approach to state level GIS in Tamil Nadu at Paramakudi taluk of Ramanathapuarm district using GIS. Large scale maps (1:5000) for above 93 villages have been vectorised and developed into a decision support system with decision support shell features.

It integrates the data available with various district level authorities includes District Education Officer, Joint Director of Medical, District Statistical Officer, Taluk Supply Officer, District Rural Development Agency (DRDA), etc. Decision Support System features have been available in the following modes.

- Selective Data Display
- Attribute Display
- Query Builder
- Model Builder and
- History of Development Schemes.

This has become an operational user interface for decision making.

Application Areas

The emergence of GIS and its variety of applications in many areas are indicated as below.

- Address Matching
- Bio-diversity evaluation
- Catchment Area Definition
- Census Analysis
- Conveyancing
- Crime Pattern Analysis
- Crowd Control
- Eco System Management
- Emergency Planning
- Epidemiological Studies

- Estate Management
- Facility Location
- Flood modeling
- Forest Management
- Geological Modeling
- Ground water Protection
- Highway Maintenance
- Hydrological Modeling
- Land Use Planning
- Local Government
- Marketing
- Pollution Monitoring
- Risk Management
- Soil Erosion Studies
- Tourism Management
- Water Management

CONCLUSION

GIS originated only in the middle of 1960s and it has become a phenomenon only in the 1980s. Expansion in usage of GIS has begun in early 1980s due to combined effects of developments in software, cost effectiveness of hardware. At the same time, remote sensing also originated at similar point of time. In this context, it is felt that strong theoretical framework is very much essential for greater generality of GIS. Even though the level of theoretical development in GIS is much higher in 1980s, it required close liaison between basic research and GIS design. Therefore, GIS and its allied fields (e.g. Remote Sensing, GPS) add up to the makings of the science of spatial information. Significance of GIS as a scientific tool and its value is expanding. Understanding the world around us depends on comprehending the significance of spatial processes.

Acknowledgement

The assistance rendered by Mr. M. Murugiah, and Mr. B. Rajeshkanna, Research Scholars from Department of Futures Studies, Madurai Kamaraj University, Madurai is acknowledged.

REFERENCES

Anon., (1992).Macro-level Urban Information System: A GIS Case Study for BMR, Ahmedabad: Space Applications Centre (SAC) and Bombay Metropolitan Region Development Authority (BMRDA) - SAC/RSA/NRIS-URIS/PR-18/March 1992.

Burroughs, P. A., (1988). Principles of Geographical Information Systems for Land Resources Assessment. New York: Clarendon Press.

NCGIA, (1990). NCGIA Core Curriculum: Issues in GIS. A National Center for Geographic Information and Analysis. California: University of California, Santa Barbara.

Patkar, V.N., and D. Sampath Kumar, (1999). A Facelift to Mumbai. GIS @ Development, Jan - Feb. 1999

Rao, M., (1993). GIS Database Design. Ahmedabad: Space Applications Centre.

Rao, M., (1996). Geographical Information Systems : State of Art. Tiruchirappalli: Bharathidasan University.

Vasudevan, S., (1995). Comparative Evaluation of Thematic, Numerical and GIS Modelling: An Analysis Through Hard Rock Aquifer Data of Vellar Basin, Central Tamil Nadu, India. Unpublished Ph.D Thesis, Tiruchirappalli: Bharathidasan University.

APPLICATIONS OF GEOSPATIAL TECHNOLOGY
A. Ganesh (Ed.), (2006) : pp. 43-51
Satish Serial Publishing House, New Delhi
E-mail: ssp@ndf.vsnl.net.in, hkjain1975@yahoo.com
website: www.satishserial.co

3

Choosing the Right Digital Terrain Model

R. NARAYANAKUMAR AND K. KUMARASWAMY

Models are the simplified forms of reality, generally created for better understanding of the phenomenon or object under investigation. When someone views an environment they simplify the inherent complexity of it by abstracting key features to create a 'model' of the area. This cognitive exercise is influenced by the cultural norms of the observer and the purpose of the study. Geographical data models are the formalized equivalents of the conceptual models used by people to perceive geographical phenomena. In geographical models, they formalize how space is discretized into parts for analysis and communication and assume that phenomena can be uniquely identified, that attributes can be measured or specified and that geographical coordinates can be registered. Thus obviously, there are two approaches for modeling geographic data, viz., modeling discrete objects such as houses, land parcels etc., and making continues phenomena such as elevation, rain fall, etc. The prior one is known as entity approach, the latter tessellation.

Most anthropogenic phenomena such as, houses, bore wells, administrative units, roads, cables, pipelines, agricultural fields, can be handled best using entity approach. With continuous field data, although the variation of attributes such as elevation, air pressure, temperature, or soil is assumed to be continuous, and the variation is generally too complex. Thus it is necessary to divide geographical space into discrete spatial units such as raster pixels, thiessen polygons and triangular irregular nets. The resulting tessellation is taken as a reasonable approximation of reality. Both the entity and tessellation models assume that the phenomena can be specified exactly in terms of both their attributes and spatial position (Burrough and McDonnell, 1998). But in practice, there are some situations where these data models are some times accepted as reliable representations of reality, but there are many other circumstances where uncertainties force one to choose the one or the other approach, pragmatically. This article is about to shed light on these aspects.

Geographic Data Models and GIS

Geographic Information System is useful in creating virtual models of features or objects present over the Earth surface. The heart of any GIS is the data model, which is a set of constructs for representing objects and processes in the digital environment of the computer (Longley; et.al., 2001). The Earth surface features may be discrete or continuous. Discrete features are individually distinguishable separate entities, such as buildings and roads. Continuous features exist spatially between observations such as precipitation and elevation. In GIS there are two basic data models used to represent spatial features, viz., vector data model and raster data model (Chang Kang-tsung, 2002). The raster data model uses an array of cells, or pixels to represent generally continuous features. The vector data model on the other hand is closely linked with the discrete object view. To date, the vector data model has been widely implemented in GIS. This is because of the precise nature of its representation method, its storage efficiency, the quality of its cartographic output and the availability of functional tools for operations like map projection, overlay and analysis (Longley, et.al., 2001).

Digital Terrain Model and GIS

Digital Terrain Model (DTM) is used to describe a digital data set which is used to model a topographic surface i.e. a surface representing height data. To model a surface accurately it would be necessary to store an infinite number of observations. Since this is impossible, a surface model approximates a continuous surface using a finite number of observations. DTMs are created from a series of either regularly or irregularly spaced (x,y,z) data points (where x and y are the horizontal co-ordinates and z is the vertical or height co-ordinate).

DTM and DEM

There are numerous ways of modeling the spatial form of surface topography. Digital representations are referred to as either Digital

Elevation Models (DEMs) or Digital Terrain Models (DTMs). A DEM is an ordered array of numbers that represent the spatial distribution of elevation above some arbitrary datum in a landscape. DTMs are ordered arrays of numbers that represent the spatial distribution of terrain attributes. DEMs are, therefore, a subset of DTMs (Huggett Richard and Cheesman Jo, 2002)

Significance of DTM

The large number of digital elevation sources and terrain analysis tools enables researchers and operators in environmental science, agriculture, hydrology, biology and engineering to compute terrain-dependent variables and indices easier than ever before. They extend from slope and aspect analyses using a variety of algorithms to the estimation of more sophisticated terrain attributes such as the drainage network or the topographic wetness index. By using one or several terrain variables and indices, a catchment area can be characterized in terms of geomorphology, stream network patterns or landform classification. Subsequently ecosystem modeling, such as modeling the distribution of energy, water, sediments, nutrient, and pollutants depends on realistic terrain representation (Kienzle W Stefen, 2002).

Creating a DTM

Like data model, there are two different approaches available to create DTMs, viz., raster approach and vector approach. In raster GIS, a DTM is simply a grid of height values in which each cell contains a single value representative of the height of the terrain that is covered by that cell. The accuracy of this approach depends on the terrain complexity and resolution of the grid. In Vector GIS a TIN (Triangulated Irregular Network) is used to create a DTM from either regular or irregular height data. TIN is more advanced, more complex and more common form of DTM (Heywood, et.al, 2002)

TIN offers many advantages for surface analysis. First, they incorporate the original sample points, providing a useful check on the accuracy of the model. Second, the variable density of triangles means that a TIN is an efficient way of storing surface representations. Third, the data structure makes it easy to calculate elevation, slope, aspect and line-of-sight between points. The combination of these factors has led to the widespread use of the TIN data structure in applications such as volumetric calculations for roadway design, drainage studies for land development and visualization of urban forms (Longley, et.al., 2001).

Structure of DTM

There are two major structures of DTM , viz., Triangulated Irregular Networks and regular raster girds. Triangulated Irregular Networks sample surface-specific points, such as ridges, peaks, and breaks in slope, the regular network of points being stored as sets of x, y, and z co-ordinates each with pointers to their neighbors in the net. TINs are made up of facets, which are triangular planes created by joining three adjacent points. The triangles create a continuous, connected surface based upon Delauney triangulation. Regular raster grid method may use a regularly spaced square, or a triangular or rectangular gird. The choice of cell size or data resolution depends upon the application and the study is being examined.

Raster Grids

Rasters represent surfaces as a regular grid of uniformly spaced locations with z -values. One can estimate a surface value for any location by interpolating z - values among immediate grid points. The resolution of the grid - the width and height of cells - determines the precision of the raster representation. Rasters support a rich set of spatial analysis such as spatial coincidence, proximity, dispersion, and least cost paths, which can be performed rapidly.

Triangulated Irregular Networks (TINs)

TINs represent surfaces as contiguous non-overlapping triangular faces. One can estimate surface value for any location by simple or polynomial interpolation of elevations in a triangle. Because elevations are irregularly sampled in a TIN, a variable point density to areas can be applied, where the terrain changes sharply, yielding an efficient and accurate surface model. A TIN preserves the precise location and shape of surface features. Areal features such as lakes and islands are represented by a closed set of triangle edges. Linear features such as ridges are represented by a connected set of triangle edges. Mountain peaks are represented by a triangle node. TINs support a variety of surface analyses such as calculating elevation, slope and aspect, performing volume calculations and creating profiles on alignments.

Square-grided DEMs have become the most common format of DEM in use, owing to their ease of computer representation and manipulation. Because of the simplicity of grided DEMs and the ease with which this form of data can be handled within a computer, in particular within GIS software packages implementing raster functionality, they are frequently used within spatial analysis and process modeling of physical phenomena. Grided DEMs, however, have a number of disadvantages, they are as follows:

1. Data redundancy in areas of uniform terrain
2. Difficult to handle abrupt changes or differing complexity in elevation due to resolution, so that important details of land surface in complex regions are often missed
3. The resolution of the grid influences storage requirements, computational efficiency and results obtained.

Using a TIN structure often offers a more flexible and efficient alternative to overcome many of the above said problems associated with raster girds. In particular data redundancy could be significantly eliminated in TINs. The advantage of a TIN is that it allows extra information to be stored for areas of complex terrain without having to store additional redundant information for flatter, more homogeneous regions (Huggett Richard and Cheesman Jo, 2002)

Choosing the Right DTM

Which form of DTM, Raster representation or TIN representation of the surface, is one of the problems an analyst generally faces before going for any spatial analysis involving an elevation model. Choosing the right form of DTM involves the considerations, like, the purpose of the study, data source and interpolation technique. The purposes of the study largely determine the kind of the elevation surface to be created. There are a variety of purposes for which elevation models are created. In the table below a variety of choices/purposes and suitable form of DTM for a particular choice/purpose are listed.

DTMs may be derived from a number of data sources. These include contour and spot height information found on topographic maps, stereoscopic aerial photography, satellite images and field surveys. With spot heights, elevation models could be generated through interpolation. If there is an elevation surface available that could be directly used to generate DTM. When contours, digitized from topographic maps, if available, constructing TIN is easy.

The nature, accuracy and visualization of the DTM are highly affected by the kind of the interpolation technique applied to the raster elevation surface. Plate 1 has two DEMs derived from two different interpolation technique, viz., Spline and Inverse Distance Weighted,

The Choice/ Pupose	Raster Representation of a Surface	TIN Representation of a Surface
Acuracy of Surface Model	The Precision of a raster surface modle is determined by the cell dimensions. To increases the accuracy of a raster surface model, the entire raster must be resampled at a higher resolution.	A TIN Surface model has a variable point density that varies on the degree of change of slope. To make TIN more accurate, additional mass points, break lines and polygons can be added.
Reliability of Surface Features	Rasters sample the z-value of surface features on a regular grid. Features such as peaks and ridges cannot be located to a position more accurate than the grid resolution.	TINs are desinged to capture and represent surface features such as streams, ridges and peaks. These features are stored with precise coordinates and slope discontinuities such as ridges are modeled with break line.
Surface Analysis	Spatial coincidence, proximity, dispersion, least-cost path.	Elevation, slope, aspect calculations, contour derivations from surface volume profiles on alignments, line-of-sight analysis.
Scale of the Study	Rasters are appropriate for small -scale mapping applications, where positional acuracy is not paramount and where surface features do not need to be charactrized exactly.	TINs are well suited for large-scale mapping applications, where positional acuracy and shapes of surface features are important.
Sample Applications	Small-scale surface display and modeling. Modeling of pollutant dispersion, identification of watershed basins, hydrologic analysis of flood zones	Volumetric calculations for roadway design, drainage studies for land development, generation of high-quality elevation contour, perspective displays of buildings on a landscape

popularly known as IDW, using the same spot heights. Plate.2 shows a TIN surface derived from digitized contour. By comparing the three DTMs with the satellite image of the study area (Plate.2) the two DEMs derived from spot heights are not true to reality. However, the TIN surface derived from the contours digitized from the topographic map is very much near to reality.

Conclusion

TIN surfaces derived from digitized contours are highly reliable, if one goes for deriving terrain dependent variables like slope, aspect,

Plate 1: **Digital Elevation Dodel Derived From Spot Heights**

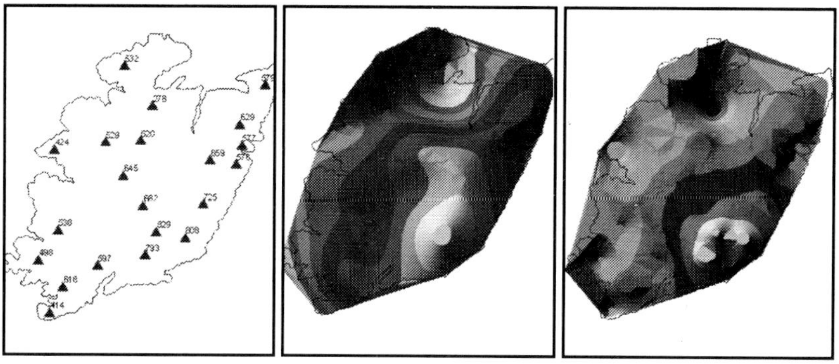

Spot Heighta | DEM Derived Form Spllne Interpolation | DEM Derlved From IDW Interpolation

Plate 2: **Digital Elevation Model Derived From Digitized contour**

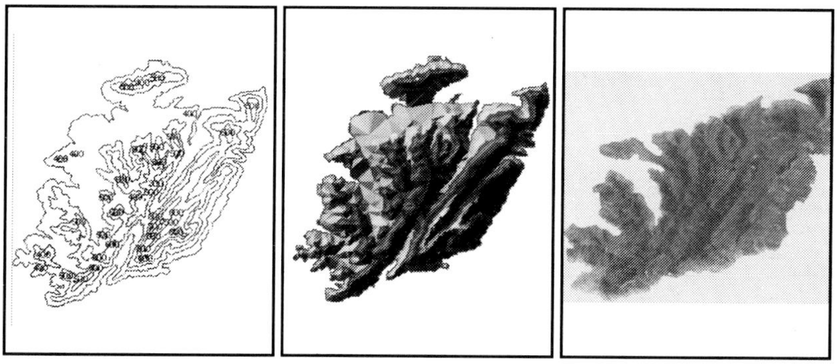

Digitized Contours | Digital Elvation Model | Landsat TM Image of the Study Area

curvature, etc. For some area where topographic maps are not available, the only option would be generating DEM with available spot heights. However, with the recent developments in photogrammetry and techniques like Lidar remote sensing, it is possible to generate precise spot heights. Further the density of the spot heights has a very high influence in the generation of accurate DEMs.

References

Burrough A. P., and M. A. Rachael, (1998). Principles of Geographical Information Systems. Oxford: Oxford University Press.

Change Kang-tsung, (2002).Geographic Information Systems. New York: Mc Graw Hill.

Heywood Ian, Cornelius Sarah and Carver Steve, (2002). An Introduction to Geographical Information System. Third Indian Reprint. New Delhi: Pearson Education, Ltd.

Huggett Richard and Cheesman Jo, (2002).Topography and the Environment. Essex: Pearson Education Limited.

Kienzle W Stefen, (year?). The Effect of Grid Cell Size on Major Terrain Derivatives. Canada: University of Lethbridge.

Longley A. Paul, Goodchild F. Michael, Maguire J David and Rhind W. David, (2001). Geographic Information System and Science. West Sussex: John Wiley and Sons, Ltd.

Zeiler Michael, (1999). Modeling Our World – The ESRI Guide to Geodatabase Design. Redlands, California: Environmental Systems Research Institute, Inc.

APPLICATIONS OF GEOSPATIAL TECHNOLOGY
A. Ganesh (Ed.), (2006) : pp. 51-63
Satish Serial Publishing House, New Delhi
E-mail: ssp@ndf.vsnl.net.in, hkjain1975@yahoo.com
website: www.satishserial.co

4

Spatial Data Exploration and Geostatistical Interpolation

K. BANUKUMAR

A major challenge facing most GIS modelers is to generate the most accurate surface possible from existing sample data as well as to characterize the error and variability of the predicted surface. Newly generated surfaces are used in GIS modeling and analysis as well as inhree-dimensional visualization. Understanding the quality of data can greatly improve the utility and purpose of GIS modeling. Geostatistics, in its original usage, referred to statistics of the Earth. Geostatistics is widely used in many fields and comprises a branch of spatial statistics. Originally in spatial statistics, geostatistics is synonymous with kriging, which is a statistical version of interpolation. One of the essential features of geostatistics is that the phenomenon being studied, takes values, not necessarily measured, everywhere within the study area, for example, the amount of rainfall in a region, or the concentration of CO_2 in the atmosphere etc.

Geostatistical analyst uses sample points taken at different locations in a landscape and predicts interpolation results into continuous surface. The sample points are measurements of any phenomenon such as water quality, pollution source, magnitude of earthquake or elevation heights etc. Geostatistical analyst provides two groups of interpolation techniques namely deterministic and geostatistical techniques. Deterministic technique uses mathematical functions for interpolation whereas in geostatistical function, it incorporates statistical inferences along with mathematical modeling to predict a surface. Prior to any method of interpolation, the data must be normalized. The skewed data can be normalized thorough transformation option after data exploration. Whatever the interpolation techniques, the accuracy of the prediction depends on the efficient usage of existing data and it can be achieved thorough investigation of the sample data. Generally a data set can be examined in its distribution, trends if any persist and spatial relationship among the data within the data set. With this brief introduction the present study investigates and emphasizes on the different means of data exploration methods using Geostatistical Analyst extension of ArcGIS 9.1.

Brief descriptions about transformation, different interpolation methods such as Inverse Distance Weighted, Spline and different types of Kriging methods have also discussed. Post monsoon mean water level data for Tiruchirappalli District, Tamil Nadu, India have been taken as sample dataset for the data exploration analysis. The geostatistical analysis platform offers different types data exploration methods including Histogram, QQ Plot, Trend Analysis and Semivariogram/Covariance Cloud investigation. Voronoi Map and Cross Covariance investigations also can use for data exploration which is offered by the geostatistical method of interpolation.

Deterministic Type of Interpolation

Generally speaking, things that are closer together tend to be more alike than things that are farther apart. This is a fundamental geographic principle. Suppose you are a town planner and you need to build a scenic park in your town. You have several candidate sites and you may want to model their viewsheds at each location. This will require a more detailed elevation surface dataset for your study area. Suppose you have pre-existing elevation data for 1,000 locations throughout the town. You can use this to build a new elevation surface. When trying to build the elevation surface, you can assume that the sample values closest to the prediction location will be similar. But how many sample locations should you consider? And should all of the sample values be considered equally? As you move farther away from the prediction location, the influence of the points will decrease. Considering a point too far away may actually be detrimental because the point may be located in an area that is dramatically different from the prediction location.

One solution is to consider enough points to give a good prediction, but few enough points to be practical. The number will vary with the amount and distribution of the sample points and the character of the surface. If the elevation samples are relatively evenly distributed and the surface characteristics do not change significantly across your landscape, you can predict surface values from nearby points with reasonable accuracy. To account for the distance relationship, the values of closer points are usually weighted more heavily than those farther away. This is the basis for the Inverse Distance Weighted (IDW) interpolation technique. As its name implies, the weight of a value decreases as the distance increases from the prediction location.

Geostatistical Interpolation

Geostatistical techniques rely on both statistical and mathematical methods, which can be used to create surfaces and assess the uncertainty of the predictions. After exploring the data, you can create a variety of output map types including prediction, error of prediction, probability and quantile maps using many variants of kriging and cokriging algorithms and associated tools. Generating a continuous surface used to represent a particular attribute is a key capability required in most of the GIS applications. Perhaps the most commonly used surface type is a digital elevation model of terrain. These data sets are readily available at small scales for various parts of the world. However, just about any measure taken at locations across a landscape, subsurface, or atmosphere can be used to generate a continuous surface. The kriging process associates some probability with your predictions; that is, the values are not perfectly predictable from a statistical model alone.

Data Exploration

Data exploration allows us to examine the data in different ways. Before creating a surface, it is necessary to gain a deeper understanding of the phenomena of our investigation so that it can make better decisions on issues relating to the data. The geostatistical environment is composed of a series of tools, each allowing a view into the data. Each view can be manipulated and explored allowing different insights about the data. Each view is interconnected with all other views as well as with ArcMap. That is, if a bar is selected in the histogram, the points comprising the bar are also selected on the QQ plot.

Histogram Examination

The Histogram tool provides univariate (one-variable) description of the data. The tool displays the frequency distribution for the dataset of interest and calculates summary statistics. The frequency distribution is a bar graph that displays how often observed values fall within certain intervals or classes. You specify the number of classes of equal width that should be used in the histogram. The relative proportion of data that falls in each class is represented by the height of each bar. The important features of a distribution can be summarized by statistics that describe its location, spread, and shape measures.

In histogram the sample value are shown in Y axis as a frequency and number of samples as shown in X axis in square of original measurements.

Measures of Location: Measures of location provide an idea of where the center and other parts of the distribution lie. The mean is the arithmetic average of the data. The mean provides a measure of the center of the distribution. The median value corresponds to a cumulative proportion of 0.5. If the data is arranged in increasing order, 50 per cent of the values would lie below the median and 50 per cent of the values would lie above the median. The median provides another measure of the center of the distribution. The 1st and 3rd quartiles correspond to the cumulative proportion of 0.25 and 0.75, respectively. If the data is arranged in increasing order, 25 per cent of the values would lie below the first quartile, and 25 per cent of the values would lie above the third quartile. The 1st and 3rd quartiles are special cases of quantiles. The quantiles are calculated as follows:

$$\text{quantile} = (i) - 0.5 / N$$

where (i) is the ith rank of the ordered data values.

Measures of Spread: The spread of points around the mean value is another characteristic of the displayed frequency distribution. The variance of the data is the average squared deviation of all values from the mean. The units are the square of the units of the original measurements and because it involves squared differences, the calculated variance is sensitive to unusually high or low values. The standard deviation is the square root of the variance. It describes the spread of the data about the mean in the same units as the original measurements. The smaller the variance and standard deviation, the tighter the cluster of measurements about the mean value. In the given ground water data standard deviation value shows the skewed distribution of data and six controlled wells are identified higher correlation with its mean.

Measures of shape: The frequency distribution is also characterized by its shape. The coefficient of skewness is a measure of the symmetry of a distribution. For symmetric distributions, the coefficient of skewness is zero. If a distribution has a long right tail like present condition of large values, it is positively skewed and if it has a long left tail of small values, it is negatively skewed.

The mean (6.0) is larger than the median (5.0) for positively skewed distributions and vice versa for negatively skewed distributions. The kurtosis is based on the size of the tails of a distribution and provides a measure of how likely the distribution will produce outliers. The kurtosis of a normal distribution is 3. Distributions with relatively thick tails are "leptokurtic" and have kurtosis greater than 3 as in present data here Kurtosis observed as 4.1. Distributions with relatively thin tails are "platykurtic" and have a kurtosis less than 3.

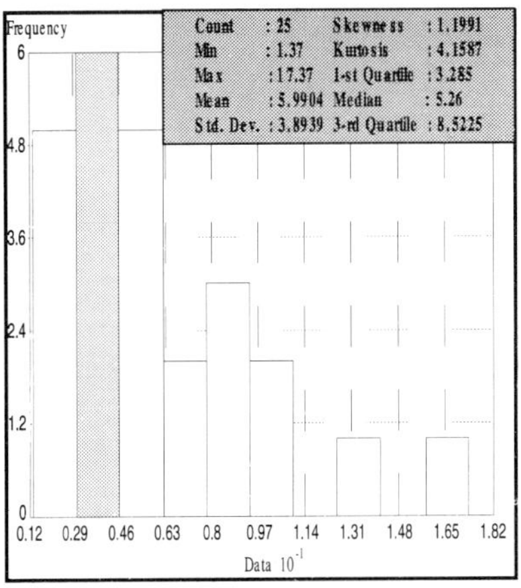

Fig. 1(a)

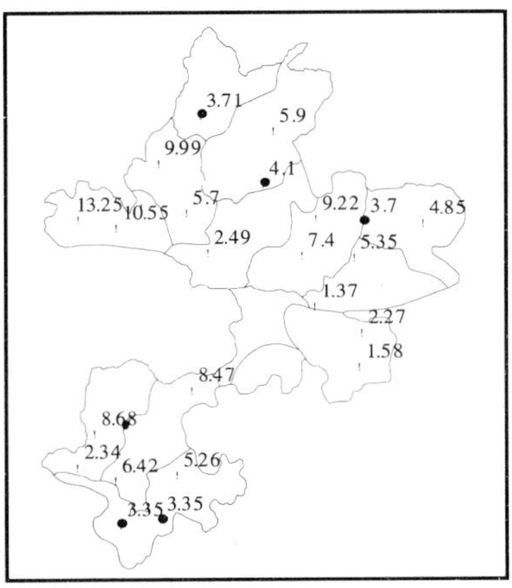

Fig. 1(b)

Figure 1 (a) shows all the statistical parameter of the given ground water level data and it provides an interactive data selection between the graph and the map shown in Figure 1 (b).

QQ Plot Investigation

A scatter plot where the quantiles of two distributions are plotted against each other. A cumulative distribution for the data is produced by ordering the data and producing a graph of the ordered values versus cumulative distribution values calculated as (i– 0.5) / n for the i th ordered value out of n total values.

Normal QQ Plot is created by plotting data values with the value of a standard normal where their cumulative distributions are equal shown in Figure 2 below.

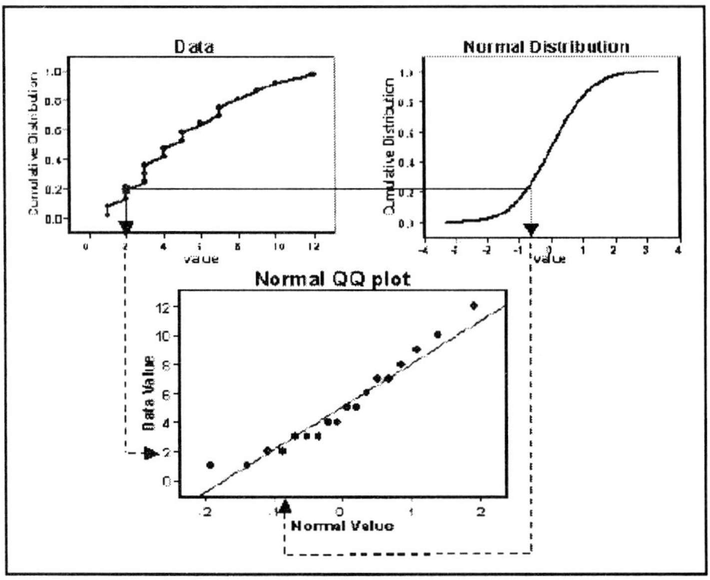

Fig. 2

For the cumulative distribution, the median value splits the data into two halves, while quartiles split the data into quarters, deciles split the data into tenths and percentiles split the data into hundredths. Quantiles are an extension of this idea. They split the data into any fraction. For example, 33 per cent of the data will lie below the 0.33 quantile. In Figure 3 the QQ plot shows the deviation of data from mean (in X axis) and percentage of data (in quantile) under this range and one can understand that more than 55 percentage of data are below mean and it also exhibits how the overall data value deviate from the normal distribution with reference to the straight line.

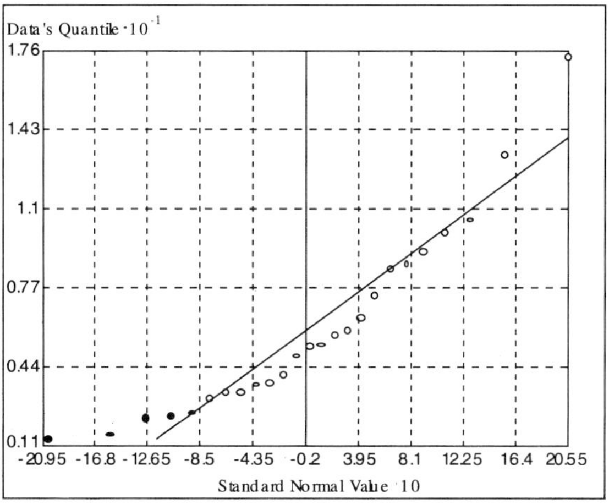

Fig. 3

Trend Analysis

A surface may be made up of two main component a fixed global trend and random short-range variation determined by the data quality and quantity. Random short-range variation sometimes referred to as random error. After looking for global trends, if identified, global trend exists in the data, then decide to model it either by using a deterministic method or a geostatistical method to create a surface usually depends on the objective. In order to model just the global trend it will create a smooth surface a global or local polynomial interpolation method can enough to create a final surface.

However, incorporating the trend in a geostatistical method has been carried out by means removing the global trend with justifying it, and then model the remaining component as random short-range variation. The elimination of the global trend in a geostatistical method will be model the random short-range variation in the residuals. However, before making an actual prediction, the trend will be automatically added back so that reasonable results can be obtained.

The trend tool raises the points above a plot of the study site to the height of the values of the attribute of interest in a three-dimensional plot of the study area. The points are then projected in two directions by default, north and west onto planes that are perpendicular to the map plane. A polynomial curve fit to each projection. The entire map surface can be rotated in any direction, which also changes the direction represented by the projected planes. If the curve through the projected points is flat, no trend exists, as is shown with the blue line in the right projected plane and if there is a definite pattern to the polynomial, such as a downward curving model as shown with the green line in the back left projected plane this suggests a trend in the data shown in Figure 4 (a).

If you rotate the sample points 30 degrees in order perceive direction shown Figure 4 (b) and it demonstrates a gently upward 'U' shaped curve suggests that a second-order polynomial can fit to the data. Through the refinement allowed in the Trend Analysis tool, the true direction of the trend can be identified. In this case, its strongest influence is from the southeast to the northwest.

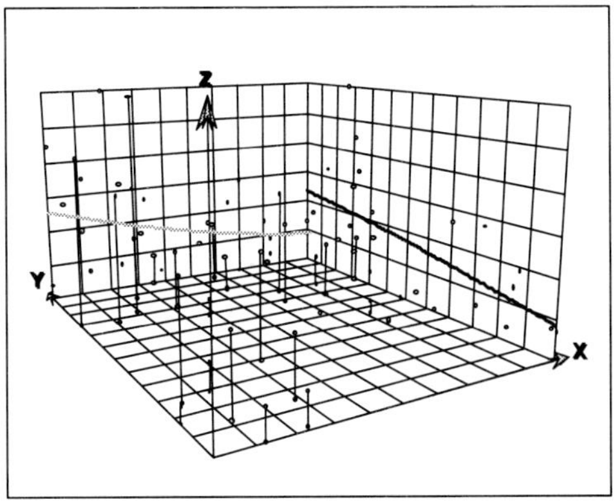

Fig. 4 (a)

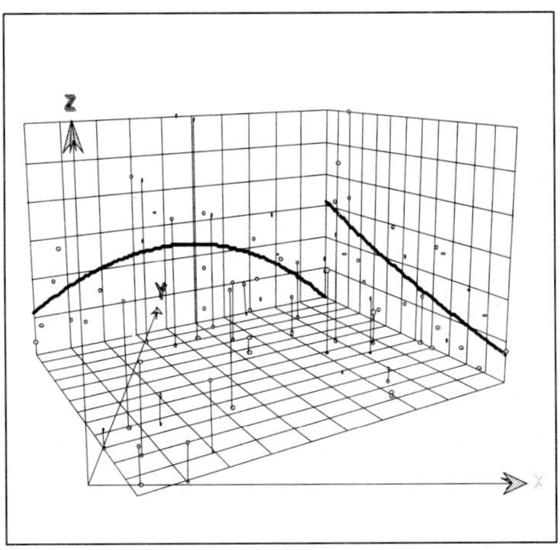

Fig. 4 (b)

Semivariogram/Covariance Cloud

The Semivariogram/Covariance cloud shows the empirical semivariogram and covariance for all pairs of locations within a dataset, and plots them as a function of the distance between the two locations. This graph can be used to examine the local characteristics of spatial autocorrelation within a dataset and look for local outliers.

Autocorrelation: Correlation is usually thought of as the tendency for two types of variables to be related. For example, the stock market tends to make positive changes with lower interest rates, so it's said that they are negatively correlated. However, the stock market is positively autocorrelated, which means it has correlation within itself. In the stock market, two values will tend to be more similar if they are only one day apart, as opposed to being one year apart. This is related to a basic principle of geography—things closer together tend to be more similar than those that are farther apart. The rate at which the correlation decays can be expressed as a function of distance.

In Figure 5 (a) each dot is the empirical semivariogram, the difference squared is plotted in Y axis relative to the separation distance plotted in X axis, between a pair of locations in the dataset. You can brush dots and see the linked pairs in ArcMap. Plotting each pair of locations quickly becomes unmanageable.

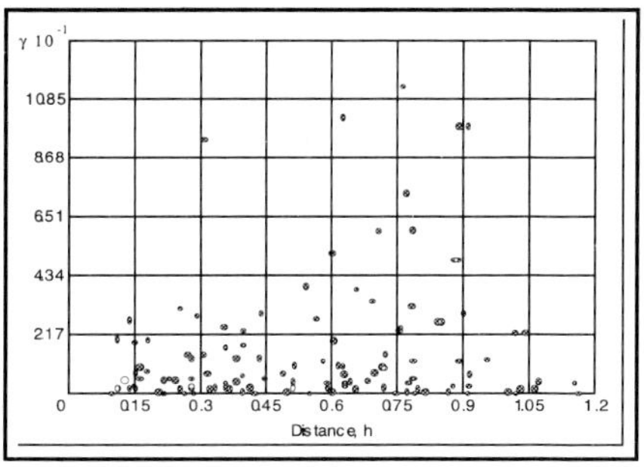

Fig. 5(a)

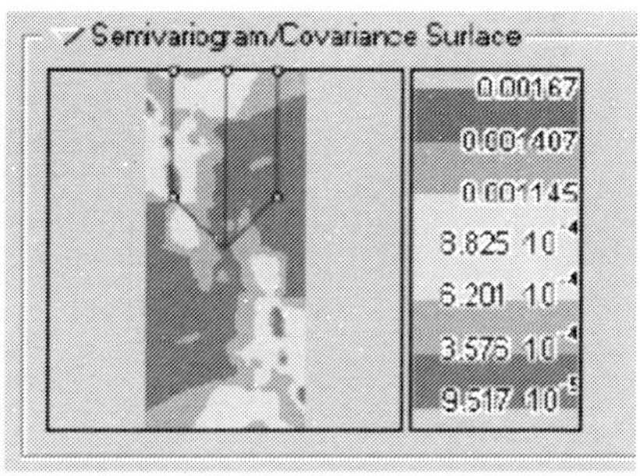

Fig. 5(b)

To reduce the number of points in the empirical semivariogram, the pairs of locations will be grouped based on their distance from one another. This grouping process is known as binning. Binning is a two-stage process. First, form pairs of points, this process continue for all possible pairs. You can see that in the pairing process the number of pairs increases rapidly with the addition of each location. This is why, for each bin, only the average distance and semivariance for all the pairs in that bin are plotted as a single point on the semivariogram.

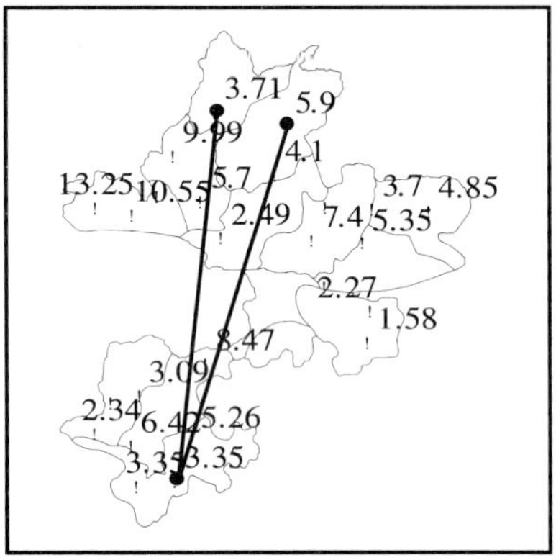

Fig 5(c)

In the second stage of the binning process, pairs are grouped based on common distances and directions. Imagine a graph in which each point has a common origin. This property makes the empirical semivariogram symmetric. For each bin, you form the squared difference from the values for all pairs of locations that are linked, and these are then averaged and multiplied by 0.5 to give one empirical semivariogram value per bin. In Geostatistical Analyst, you can control the lag size and number of bins.

The empirical semivariogram value in each bin is color coded and is called the semivariogram surface. The selection of a lag size has important effects on the empirical semivariogram. For example, if the lag size is too large, short-range autocorrelation may be masked. If the lag size is too small, there may be many empty bins, and sample sizes within bins will be too small to get representative "averages" for bins. However, data usually acquired or available in an irregular or random sampling scheme,the selection of a suitable lag size is not so straightforward. A rule of thumb is to multiply the lag size by the number of lags, which should be about half of the largest distance among all points. Such way in present study the lag size is maintained 0.12 with 10 lags in order to incorporate all the data pairs of semivariogram.

A Semivariogram Surface with Search Direction capabilities is shown in Figure 5 (b). The values in the semivariogram cloud are put into bins based on the direction and distance between a pair of locations. By adjusting band (lag) width can determine the pairs to be used for semivariogram construction with certain distances and direction apart. Band width and distance of pair has inverse relationship. In the present study 3 km width is used to incorporate the samples within this distance which may have higher influence and can be used for surface prediction. Figure 5 (c) demonstrates what will be the search direction in between these connected 3 values and we can decide whether the data has to be included or not according to the predicted surface.

Findings from Data Exploration

The histogram investigation of the ground water level data for Tiruchirappalli district reveals that all the basic statistical parameters are calculated such as minimum (1.47 m) maximum (17.4 m), mean (5.9 m), standard deviation (3.8 m), skew (1.2), kurtosis (4.1) and median (5.2 m). In the given ground water data standard deviation value shows the skewed distribution and six controlled wells are identified higher correlation with its mean. The distribution of data has a long right tail and it is considered to be positively skewed. Usually the mean (6.0) is larger than the median (5.0) for positively skewed distributions have also observed from this histogram exploration. The QQ plot data exploration shows the deviation of data from mean and percentage of data under this range. One can understand that more than 55 percentages of data are below mean and it also exhibits how the overall data value deviate from the normal distribution with reference to the straight line of QQ plot.

It is observed from the trend surface data exploration that the curve through the projected points with a downward curving model as shown in the line in the back left projected plane suggests a trend in the ground water level data. The rotation of the sample points in 30 degrees perceives direction gently upward 'U' shaped curve, which suggests a second-order polynomial can fit to the data for interpolation. There is a strongest influence from the southeast to the northwest which may be caused due to hydro geological setting of the area. With reference to semivariogram - covariance cloud, it is clear that the search direction, number of pairs are in the use, its distances interactively with ArcMap and the search direction, pair inclusion and exclusion and pair grouping can be modified depending upon any specific need and also can be used the error in the data set if any.

Conclusion

Geostatistical Analyst provides a number of interpolation techniques that use sample points to produce surfaces of the phenomena of interest. The interpolation techniques in Geostatistical

Analyst are divided into two main types, deterministic and geostatistical. Deterministic techniques are based on parameters that control either (i) the extent of similarity (e.g., Inverse Distance Weighted) of the values, or (ii) the degree of smoothing (e.g., Radial Basis Functions) in the surface. These techniques do not use a model of random spatial process. Geostatistics assume that at least some of the spatial variation of natural phenomena can be modeled by random processes with spatial autocorrelation.

Before using the interpolation techniques, data explored with data exploration tools and it enabling to select the appropriate parameters for interpolation. For example, when using ordinary kriging to produce a quantile map, you should examine the distribution of the data because it is assumed that the data is normally distributed. Alternatively the trend in data can be explored with these tools and with justification the trend can be eliminated in the prediction process. Data exploration through histogram, quantile plot, trend surface and semivariogram/ covariance are the key step between spatial description and spatial prediction.

However, the Geostatistical Analyst provides different functions to model the empirical semivariogram such as Circular, Spherical Tetraspherical, Pentaspherical, Exponential, Gaussian, Rational Quadratic, Hole Effect, K-Bessel, J- Bessel and Stable. The selected model influences the prediction of the unknown values, particularly when the shape of the curve near the origin differs significantly. The steeper the curve nears the origin, the more influence the closest neighbors will have on the prediction. As a result, the output surface will be less smooth. Thorough data exploration and understanding of phenomenon will help in choosing better model to interpolate the surface. The geostatistical extension also offers a validation tool in order to maintain the quality of the predicted surface. With reference to the ground water data exploration, the present study reveals that the simple circular or exponential model will not produce much useful surface and it can be attained through the combination of more than one model.

REFERENCES

Chilies, J., and P. Delfiner, (1999). Geostatistics: Modeling Spatial Uncertainity. New York: John Wiley and Sons, p. 695.

Cressie, N. 1990. The Origin of Kriging. Mathematical Geology, 22 pp. 239 -252.

Cressie, N., (1986). Kriging Non Stationary Data. Journal of the American Statistical Association 81, pp. 625-634.

ESRI, (2003) Using Arc GIS Geostatistical Analyst ESRI Manual. Redlands: ESRI.

Gandin, L.S., (1960). On Optimal Interpolation and Extrapolation of Meteorological Fields. Trudy GGO 114, pp. 75 – 89.

APPLICATIONS OF GEOSPATIAL TECHNOLOGY
A. Ganesh (Ed.), (2006) : pp. 65-77
Satish Serial Publishing House, New Delhi
E-mail: ssp@ndf.vsnl.net.in, hkjain1975@yahoo.com
website: www.satishserial.co

5

Analysis of Map Overlay Function in GIS

K. LAKSHMI

The quest for a logical way to represent the real world has been one of the general themes of science in geography especially in the field of cartography for centuries. The natural and environmental data are measured and documented in terms of topographic and thematic maps. Maps are the most efficient analog means to display spatial and non-spatial (attributes) data. The geographic data matrix proposed by Berry (1964) was a milestone in the development of scientific way of organizing and analyzing geographic information to understand the spatial distribution and differentiation in the pattern. In the data matrix, the row represents the variation of specific aspects of natural or socio-economic characters across some geographic space. The column on the other hand denotes specific location in geographic space. This data matrix is become the base for computer mapping technology. The computer-assisted cartography using GIS helps for accurate measurement, comparison, analysis, inference, exploration and generation of spatial patterns in multi-dimensions. The software of GIS and the computer hardware can be used as a central storehouse of spatial storage of data in computer in a required format. There are several theoretical and practical ideas about the analysis of spatial data with reference to space and time through GIS for understanding, representing, analyzing and planning. Map overlay in one of the best GIS technology, which creates inspirations for researchers to understand various combined elements in the spatial set up. Before the advent of GIS, using a series of transparent maps the overlays were carried out. Through the overlay functions the areas that deemed suitable and unsuitable are blocked out by a set of binary maps (yes or no) and is stacked on top of each other henceforth the delimitation was made.

Map Overlay

In GIS operations the map overlay is based on Boolean logic and Sieve mapping. Boolean logic emphasizes the importance of overlay inputs on the correctly geo-registered same co-ordinate system. Sieve mapping in GIS is the version of old technique used by land use planners to identify areas suitable or not suitable for same function. McHarg, (1969) first formalized this technique for mapping land suitability. Sieve mapping involves the process of disqualified or eliminations of factors successively on the basis of each criterion, until those remaining are found to be susceptible on all the other criteria. Boolean logic operation in GIS supports map combination in possible ways to achieve a required outcome. The fundamental operation here is merge and geometric intersection. These two operations are the development of union and intersection operations, similar to Set theory and Venn diagrams application in geographical studies.

The union is the merges of polygons form different layers depending on the similarity of their interdependent attributes. This is essentially the operation of geometric manipulation of input polygons. The geometric intersection and merges are good examples of the interchangeability of raster and vector models. This technique is normally applied to a single map made up of points, lines and areas. Map overlay of various type of coverage produces a composite coverage formed from multiple intersections of all input layers. Each area in the composite layer has multiple attributes, derived from the attributes of their base and can be assigned an overall suitability for the satisfaction of the objective of the theme.

Carver in 1991 described few steps for map over lay. They are determining the data inputs, getting data inputs, referencing or co-registration, transformation and overlaying.

1. D etermining the data inputs: This is a knowledge driven approach and includes decision of maps to overlay, the objective of the map representation, the data sets and maps to overlay and the application of Boolean /Sieve operation.
2. Getting the data: The inputs of the data are often scanned and digitized from maps with different geo-referenced co-ordinate systems. To get accurate result of overlay and to avoid the error in displacement, getting data inputs to the same co-ordinate system is essential. A grid-on-grid transformation is required to co-register the different sources of database. The nature of the

data in GIS environment may be derived from many ways. For example, the slope estimation may be either by spot heights, contours or estimated by digital elevation matrix.

3. Re ferencing or the co-registration: This is based on transformation of data. This is an important function in GIS due to the problems of co- registration, which often arises in GIS when transforming data from one set of co-ordinate into another and integrating data from several sources.

4. Transformation and overlay: A standard GIS tool kits have three operations such as translation, scaling and rotation. These three operations can be done in combine for a single transformation. Once the input maps are correctly co-registered, it is finally possible to overlay them in a raster environment; this is extremely easy, the system works across the map pixel by pixel by possible combination of conditions for the overlay. This operation can also be performed in a vector GIS environment.

Vector Based GIS

Conventionally, the topologically structured layer has been the standard data format for vector GIS. The vector GIS data processing relies on topological relationship for understanding spatial organization of geographical location attributes. The representation of spatial relationship is not simple due to numerous, various and dynamic combination of spatial phenomena.

Conditions of Map Overlay

Map overlay can only work with two polygon maps at a time. If necessity for third overlay map, the operation begins with the first two and then use the output from them with the third.

- The most common overlay operation is polygon on polygon. Each new set polygon in the output map carried attributes from both maps. These attributes differ from those of the adjacent polygons. For example polygon on polygon overlay is used for analyzing the association between land type and vegetation. Thus two aspects combination produce new identify function of land suitability for vegetation.
- The three common overlay methods of union, intersect and identify are normally in use. Union preserves all attributes from the input and overlay maps by combining the area of both maps.

Union requires both input and overlay maps be polygon maps and it is based on Boolean operation.

- Intersect is often a preferred method of overlay because any map feature on its output has attribute data from both of its imports. The input map for intersect may contain points, lines or polygon.
- The area of identity that preserves only map features that fall with in the areal extent defined by the input map. The features of the overlay map, which fall outside the areal extent of the input map, are left out of the output. The input map for identity may also contain points, lines or polygons.

Vector Based Data Processing in GIS

The major operation in the vector based GIS map overlay function is briefly outlined as follows:

Input Function

Map scanning

Digitizing in ERDAS/ESRI

Editing

Geometric registration

Shape file creation

Importing

ArcInfo / ArcView GIS

Analyzing Functions:

Non topological analysis	Topological Analysis	
1. Attribute Database Creation	Feature Based Analysis	Layer Based Analysis
2. Area calculation		
3. Statistical computation	1. Buffering	1.Reclassification
	2. Feature based overlay Analysis	2.Layer based overlay Analysis
	3.Network Analysis	

Output functions

1. Vector map composition
2. Vector – raster conversion

To understand the map overlay in vector based geographical data processing analysis the two type of non-topological and topological functional operations are outlined here.

Non-topological

The non-topological analysis is a function of spatial data display, spatial data query, attribute data query and statistical computation. The storage of descriptive data in the form of tables in the geo-relational and object - oriented models allows these data to be retrieved, updated and analyzed by means of data base query commands. The attribute data should be entered into common identifiers of polygons in the data file. The physically separate files for each map also can be linked as one file during the data processing in GIS. The summarizing the attributes in a polygon can generate the thematic maps. The statistical functional capability of GIS helps for the aggregation of attribute data for visualization, interpretation and preparation of analytical maps for planning.

Topological

The topological operation analysis consists of feature-based analysis and layer based analysis. In feature based analysis there are two operations such as buffering and feature based overlay analysis.

Buffering

Buffering is a very important function in vector based geographical data processing. Buffer is a zone with specified width surrounding a spatial feature. The process of creating a buffer is buffering. The buffer that is created around a particular spatial feature, which can be a point, a line and an area. In vector GIS operation buffering created along the river basin for the delimitation of flood plain area, quality based water zone, soil characters and elevation. In most GIS analysis buffering include uniform buffer distance for delineation. This is the most important application for environmental planning in restricted areas, transport and urban planning.

Feature Based Overlay Analysis

The aim of the feature based overlay analysis is to find the relationship between individual feature of one layer and those of another layer. One of the most commonly used feature based overlay is point in polygon matching. The objective of function is to determine whether a given point feature falls inside a polygon. The location, allocation analysis of facilities and services for community planning can be well done by this method. Even the protected species location in forest area also can be done by this method for forest resources management studies.

Layer Based Topological Analysis

Layer based topological analysis is the function of particular class of geographical data in the topologically structured format. The function is based on entire layer than an individual spatial feature. Layer based topological functions, as a rule, are computationally intensive in nature. This operation has two functions, they are reclassification and layer based overlay analysis.

Reclassification

Reclassification is simplification process, which aims to reduce the number of categories of attribute data on a layer. This process can be done by the method of non-topological conversion of data to select particular value or range of values to the attribute data and coding them according to new value. After reclassification, the polygons with some attribute values can be merged to create common boundary and remove more division. Example original soil type map can be reclassified on the basis of land suitability for cultivation.

Clay	-	unsuitable
Gravel sandy sterile soil	-	unsuitable
Sandy loam	-	suitable
Sandy clay	-	moderately suitable
Sand	-	moderately suitable

Five types of soil attributes are reclassified into three to generate new attribute table and to establish the topological relationship with geographical elements in the new thematic layer.

A GIS approach study of landslide was made by Wang and Unwin(1992) to identify landscape factors thought to be significant in land sliding. A binary map was created for each factor. The areas such that areas coded as 1 (one) are thought to be susceptible to landslides due to that factor and such those coded 0 (zero) have thought to be safe. All maps are then combined by an overlay operation to identify areas that are susceptible or not, based on all factors. The basic level of GIS operations of windowing and buffering also involved in the overlay of polygons.

Layer-based topological overlay in GIS

Burrough (1992) and Unwin (1996) recognized ten possible type of map overlay through the geometric viewpoints. The general ways in which one can combine different type the geographical objects of point, lines, areas and field in different combination was discussed by them. To evaluate the spatial attributes in a common scale, scoring techniques also discussed by them. For example rock or type of soil data may be from satellite imagery land use or from the interpretative mapping of soil survey or geology. In both cases the data land parcel is assigned in a generalization. Similarly the land use type with different properties also available in different scales. Hence to account the problems of different error and generalization in GIS, a concept of favorability function is introduced. Graeme Bonham-Carter (1995) applied a Boolean overlay to evaluate the favorability of parcels of land for some activity of process.

The evaluation of favorability 'F' can be in graduated scale of measurement such as an ordinal scale like low, medium and high, or even in ratio scale. The evaluation also can be done by coding each of the criteria used on the map as some ordinal or ratio scale. The weighting of the criteria also can be made to reflect the relative importance of the data for example by using some mathematical score function or by adding the scores. Putting all the functions occurs at each location in the input map using values at the same location in the output maps.

$$F= f\ (W_1X_1\ ----W_m\ X_m)$$

In this F is the output favorability, f represents 'same function of', and each of the input maps X_1 to X_m is weighted by an appropriate weight W_1 to W_m. This approach is called overall index maps or indexed overlay. In this operation, the values are added together to obtain a favorability score for each location. On the basis of the score, identification and delineation of problem areas, environmental risk and hazard zones and resource potential areas can be made.

Application of Overlay Model – A Case Study of Madurai District

To highlight the advantages of map overlay function in GIS for resource planning Madurai district has been selected. Madurai district lies between 9°38′ to 10° 20′ of North latitudes and from 77° 23′ to

78°16′ of East longitudes. This district has 7 taluks and 13 blocks. River Vaigai flowing from northwest to southeast direction bisects the district as north and south. The northern part of the district possesses six blocks namely Madurai West, Madurai East, Alanganallur, Vadipatti, Melur and Kottampatti. The southern part of the district consists of seven blocks such as Thiruparankundram, Thirumangalam, Kalligudi, T.Kallupatti, Sedapatti, Chellampatti and Usilampatti. These 13 blocks are treated as serially numbered polygons in the data matrix to enter all spatial data such as topographic and secondary information. For the present analysis the land suitability delineation has been taken as the main theme to validate the map overlay function in vector GIS.

The Central Survey Office district revenue map in the scale of 1:250,000 is the base map for the present analysis. At first step topography water features, and ground water potential contours are scanned and digitized in the ERDAS IMAGINE 8.4 for the geometric correction and for layer creation. Editing has been done from add and remove options. The attribute data manipulation is also done and it is imported to ArcView GIS 3.2A. All the attribute data are stored in the DBMS from the identity polygons. The required landscape attributes such as vegetation, soil type, surface water features, ground water availability, water quality characters and land cover information gathered from district hand book, PWD-Central Ground Water Board are also registered to the respective polygon for analysis.

Layer I - Physiography and Drainage

The spatial distribution of hills, ranges with the vegetation cover, river, streams, canals and tanks are registered with reference to the spatial location. Layer I is the output of geo-coded, translated attributes display of landscape with water potentials of Madurai District. On the basis of surface water available, irrigation, water table from MSL and water quality, the water potential polygons are generated as good, moderate and poor (Figure 1).

Layer II – Soil

Madurai district is predominantly covered with red soil with a combination of sand and gravel. The colour and texture of the soil is varying from deep red to thin red. The texture is fine to sterile. There is also a pocket of black soil and alluvium. The texture, type, and water retaining capacity of the soil become the base for the classification of fertility. On this basis the spatial distribution of soil characters three

Table - 1

Water Resources	Potential Score
Canal and well irrigation 125 mts. GWL from MSL Good water quality	Good
Tank and well irrigation 150-200 mts. GWL from MSL Medium water quality	Moderate
Well and Tank irrigation >200 mts. GWL from MSL Medium and poor quality	Poor

polygons are developed as high fertile, moderate fertile, and less fertile soil zones (Figure 2).

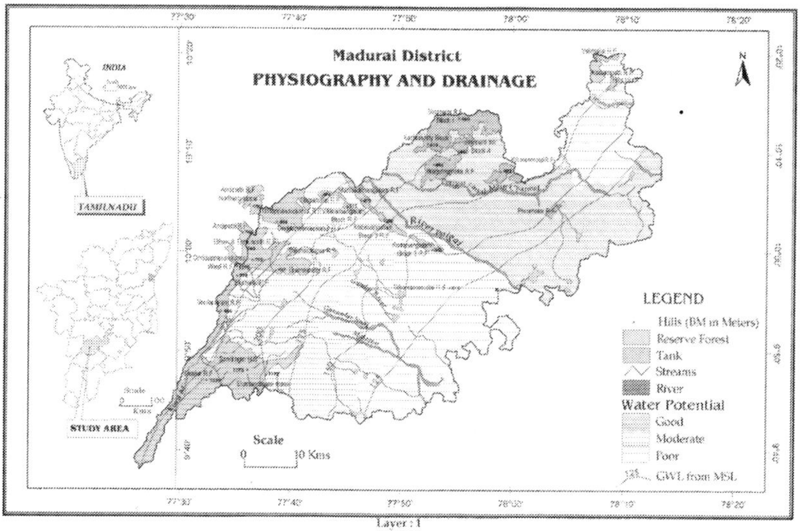

Fig. 1

Layer III – Crops

The availability of water potential and fertility of soil are the main determinants of the land suitability for cultivation. The crops cultivated in Madurai district are paddy, pulses, rain fed dry crops, sugarcane, cotton and market gardening crops. The block wise spatial distribution of crop concentration index is reclassified as three groups namely wet crops, commercial crops and dry crops and registered in the respective polygons. The output is the major crop layer of Madurai district (Figure 3).

Table -2

Soil Type	Soil suitability
Red sandy thin soil, Red sandy loam, deep red soil	High fertile
Red sandy soil, Black soil	Moderate fertile
Sterile soil & Red sandy	Low fertile

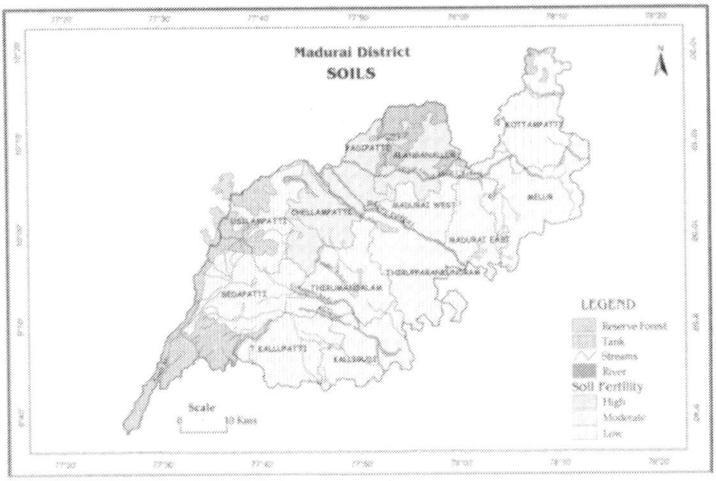

Layer 2

Fig. 2

Layer IV - First overlay layer

The overlay of water and soil (Figure 1 and 2) in the same coordinate system merges the two layer polygons. The output is integrated polygon of water and soil fertility. The overlay of three water potential on soil fertility polygons produced four polygon zones such as good water potential with high fertile soil, good fertile soil with moderate water potential, moderate soil with moderate water potential and poor water potential with less fertile soil. This output is water potential and soil fertility layer (Figure 4).

Layer V - The second overlay layer

The second overlay is output of the overlay of crop layer (Figure 3) with the first overlay layer of water potential and soil fertility layer. (Figure 4). This layer creates the merges of three factors integration. The merges of three factors produced four combination of land suitability as high, medium, low and very low.

Table - 3

Crops	Classification
Paddy, Sugarcane & Market gardening crops	Wet crops
Pulses, Cotton, Sugarcane	Commercial crops
Dry crops Pulses, Cotton	Dry crops

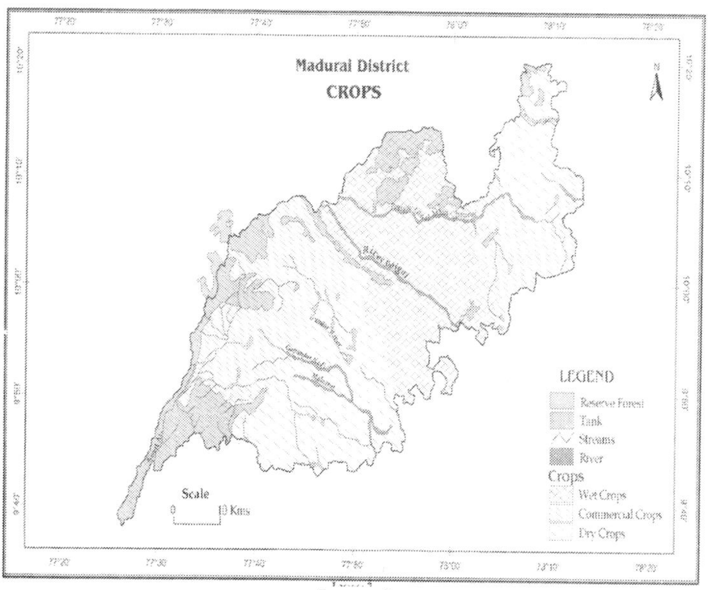

Layer 3

Fig. 3

A- The union of high potential water resources and wet crops with high and moderate fertile soil with wet crops is designated as high resource potential region for agricultural activity.

B- The union of moderate water resource potential with good fertile soil with the concentration of commercial crops is named as medium suitability zone.

C- The union of poor water resources with medium fertility of soil and the concentration of dry and commercial crops is called as low agricultural resource zone.

A- The union of poor water resources, poor fertility of soil with dry crops designated as very low resource potential region. (Figure 5).

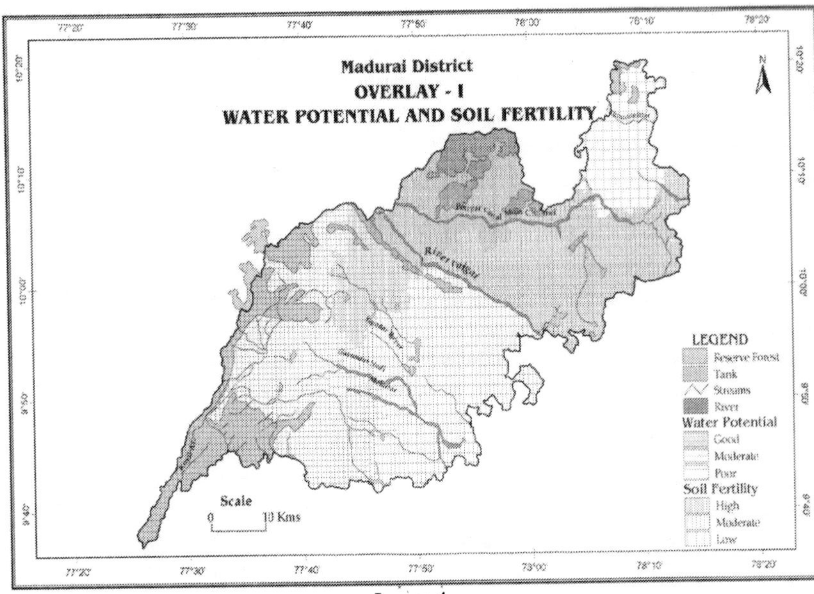

Layer 4

Fig. 4

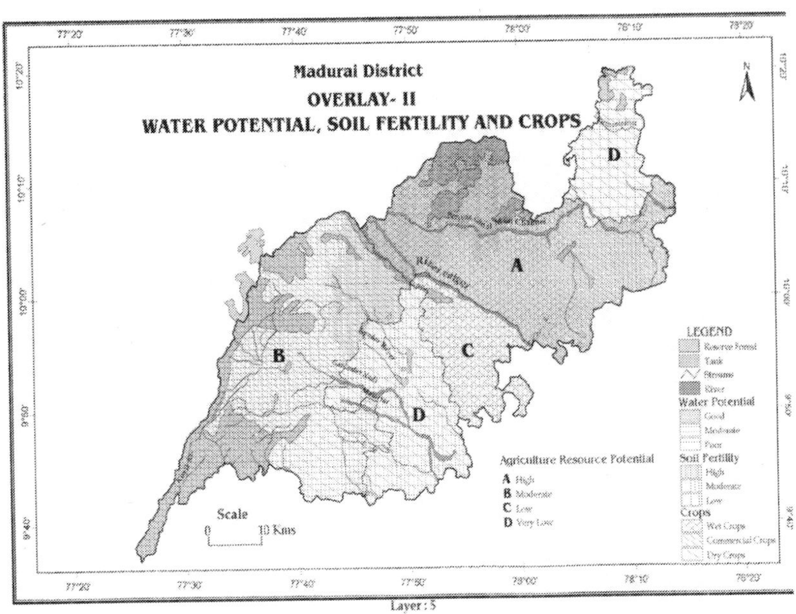

Layer 5

Fig. 5

The methodology of using overlay function through the GIS tool helps for the understanding and estimation of agriculture resource potential. It brings out that the northern part of the Madurai district except Kottampatti block has higher potential. This pattern highlights that the rich agricultural potential region is competing with the urban growth of Madurai City. This warranted the planners to restrict the urban growth at the cost of good agricultural land. This also highlights '' ıat availability of good ground water potential is also one of the restrictions for the cropping as well as urban growth.

REFERENCES

Berry, J.K. Buckley, D.J. and Ulbricht,C. (1998) Visualize Realistic Landscape: 3D Modeling helps GIS users envision Natural Resources, GIS World 1.

Burrough, P. and A.V. Fronk, (eds)(1996). Geographical Objects with Uncertain Boundaries. London: Taylor and Francis.

Burrough, P.A., (1986). Principles of Geographical Information System for land Resources Assessment, Oxford: Clarendon Press.

Burrough,P.A. and McDonnell R.A., (1998). Principles Geographical Information Systems. Oxfod: Oxford University Press.

Carver,S.J., (1991). Integrating Multi Criteria Evaluation with GIS. International Journal of Geographic Information Systems. 5 (3), pp. 321 - 339.

Ilangovan, P. and K. Lakshmi, (2000) An Atlas for Resource Development Madurai Metropolitan Area. Madurai: Nila Publisheres.

Kang-Hung Chang, (2002). Introduction to Geographic Information systems. New York: Tata McGraw Hill Edition.

LO.C.P., and Young, A.K.W., (2002). Concepts and Techniques of Geographic Information Systems. New Delhi: Prentice Hall of India Private Ltd.

McHarg, I., (1969). Design with Nature, Garden City. New York: Natural History Press.

O'Sullivan, D., and Unwin D.J., (2003). Geographic Information Analysis. New York: John Wiley and Sons. Inc.

Scholten and Unwin, (eds) (2004). Spatial Analytical Perspectives in GIS. London: Taylor and Francis, pp. 129 - 138.

Unwin, D.J., (1981). Introductory Spatial Statistics. London: Methuen.

Wang, S.Q., and D.J. Unwin, (1992). Modeling Landslide Distribution as Loess soils in China: an investigation. International Journal of Geographic nformation Systems, (5), pp. 391-405.

APPLICATIONS OF GEOSPATIAL TECHNOLOGY
A. Ganesh (Ed.), (2006) : pp. 79-111
Satish Serial Publishing House, New Delhi
E-mail: ssp@ndf.vsnl.net.in, hkjain1975@yahoo.com
website: www.satishserial.co

6

Application of Geospatial Technology for Wasteland Development

G. BHASKARAN

The problem of wasteland has become a serious issue and it has increased with the development of technology for increasing the agricultural production (Swaminathan, 1997). The natural disturbances including the man-made problems, i.e. industrialization and urbanization, contribute to increasing trend of wastelands in various ways. The requirement by the increasing human population and cattle population and also the natural disasters cause the loss of natural resources and land degradation (Hegde, 1993). According to Food and Agriculture Organization of the United Nations (FAO, 1992), the various forms of land degradation such as soil erosion, chemical poisoning, salinization and loss through building or mining about 5 to 7 million hectares from good cultivable lands.

In India, due to the growing mismanagement of the land 175 million hectares (53 per cent) of the total land area are subjected to serious environmental degradation. Nearly 80 million hectares of the degraded land are already classified as wastelands unfit for any cultivation (Yadav, 1986). The growing demand for food and fodder (70 per cent of the total population depend upon agriculture for their livelihood and livestock) turned the attention of planners, towards wasteland management and reclamation studies. To achieve this, the National Wasteland Development Board (NWDB) was created by the Government of India in 1985 with the objectives of increasing green cover, preventing good land from becoming wasteland and to promote afforestation to meet fuel wood and fodder needs of people and livestock (Dutt, 1988). In Tamil Nadu, of the total geographical extent of 130 lakh hectares include, culturable waste, permanent pastures and fallows account for 25.6 lakh hectares or 19.7 per cent of the total extent. Satellite data shows seven types of non-forest wastelands other than current fallows (11 lakhs hectares by revenue records) totaling 13.6 lakh hectares comprised mainly of upland with or without scrub (64 per cent) and lands affected by salinity or of coastal/desertic (26 per cent) nature. Not only are wastelands considerable in extent but fallows increased by 6.1 per cent in 1990 to 1999. Moreover, permanent fallows have increased to 34.3 per cent by transfer form current fallows (TAWDEV, 2001). The increasing pressure on available land resources made it difficult to meet the food and fodder requirements of human and livestock population. Bagchi and Mini Phillip (1993) have explained that the per capita availability of cultivable land has also declined from 0.48 ha (1951) to 0.18 ha (1991) and projected that it would further decline to 0.41 ha. By 2000 A.D. The alarming nature of increase in wasteland areas necessitates inventorying and monitoring them and to adopt adequate prevention and reclamation activities.

The present study attempts to understand the nature and extent of wastelands through an integrated approach using Remote Sensing (RS) and Geographic Information System (GIS) and also to understand the role of various terrain parameters in such analysis. A small micro-watershed is taken for the study since it is a basic unit for the interactive parameters such as erosion, slope, soil formation etc., which determine the genesis and type of wastelands. Moreover, the Central and State Governments currently prefer watershed approach for wasteland management for planning purposes.

Need for the Study

Reclamation of wasteland requires a detailed analysis at local level. Analysis at regional scale may give an insight about the general availability of resources and problems pertaining to wastelands. But analysis at cadastral level will give a clear picture on the intrinsic nature of wasteland problems. Hence, after regional analysis on utilization of resources and wasteland status over a period of time, it may be meaningful to select areas for cadastral details. Moreover, the true nature of the problems relating to wastelands requires consideration of not only the physical constraints but socio-economic parameters too. Such an analysis has been carried out in the present study and suggestions and recommendations have been proposed for wasteland developmental activities by counseling and interacting with the local communities for acceptability. This kind of information extraction process may be replicated in other areas in the watershed for wasteland developmental activities.

Definition of Wasteland

Stamp (1948) stated that the land which has been used previously but which has been abandoned and for which no further use has been found may be defined as a wasteland. Wasteland Survey and Reclamation Committee (1961) defined wastelands as those lands, which are either not available or left out of cultivation as fallows and cultivable waste as wastelands. National Wasteland Development Board (NWDB,1986) defines wastelands as the land which is degraded and is presently lying unutilized except as current fallow due to different constraints. Wastelands may also be defined as lands with ecological constraints and economic potentials.

On the basis of ecological constraints, wastelands may be defined as those which are ecologically unstable and whose topsoil has been nearly or completely lost and have developed toxicity in the root zone for the growth of most plants. On the basis of economic potential and actual returns, the lands that give less than 20 per cent of this economic potential are 'Wastelands'. From all the above descriptions, wasteland may be defmed as those lands which are presently unused or which are not being used to their optimum potential due to some constraints (Sinha and Dubey, 1986).

The Wasteland Survey and Reclamation Committee (1961) classifies wasteland as follows:

a. Land not available for cultivation - barren and uncultivated land.

b. Other uncultivated lands excluding fallows - culturable waste, permanent pasture and grazing lands and land under miscel laneous tree crops and groves.

c. Fallows - current as well as other types.

Wastelands include barren lands degraded to varying degrees through wind and water erosion, through deforestation, overgrazing, etc. Such lands may also be waterlogged, saline and alkaline, rocky outcrops, stony and pebbly hill slopes, sandy wastes.

Wastelands also include those lands available for cultivation but not taken up for cultivation. These lands may be abandoned after a few years owing to some reason or the other. Such lands may remain fallow for more than five years and may be covered with shrubs and grasslands and may be reserved for pasture and not included in culturable waste. In short, the wastelands are those lands that lie uncultivated for the present but could have been used previously. Such awareness on wasteland based on land capability and productivity was there from historical period in India and a brief review is given in the following section.

The National Commission on Agriculture, Government of India, published for the first time compiled from the reports of many individual government sectors. The estimate ranged from a minimum of 38.40 m.ha. to a mazimum of 187 m. ha. The estimation of extend differed from agency to agency because of the different types of classification system used by them.

Classification of Wastelands

National Wastelands Development Board (NWDB), which was formed in May 1985, created a database with a special focus on the categories and extent of wasteland toposheet-wise for critically affected districts. NWDB, with the co-operation of National Remote Sensing Agency (NRSA), Department of Space in 1984-85, took up wastelands Development Projects in146 non-sensitive districts with reference to defence and security point of view that have more than 15 per cent of wastelands. These Districts were identified by NRSA, on the basis of data available at the time using on 1: 1 Million Scale. Subsequently, the Technical Task Force Group of NWDB slightly modified the Wastelands into 13 categories as shown in Table -1.

The NRSA classification scheme forms the basis for the present wasteland analysis with little modifications based on the local problems of the study area.

The NWDB and Department of Land Resources of the Government of India, provide funds to NRSA, Hyderabad from 1986-87 to 1999-2000, for the project on on Wasteland mapping for all the districts in India. The NRSA compiled a National Atlas of Wasteland using the Landsat Thematic Mapper(TM) satellite images on 1:50,000 scale for the period 1986-1988. The agency used the Indian Remote Sensing Satellite (IRS) LISS-I, LISS-II, and LISS-III data for the year 1990-1997. The realization of problems pertaining to wasteland management have initiated the formulation of nation wide database at 1:50,000 scale by

NRSA, integrating remote sensing and GIS techniques (NRSA, 1997), which helped to device methods to solve the problems of wastelands and suggest reclamation measures.

Table 1. Wasteland clasification system

S.No	Wastelands Categories
1.	Gullied and/or ravinous land
2.	Land with or without scrub
3.	Waterlogged and marshy land
4.	Land affected by salinity/alkalinity coastal/land
5.	Shifting cultivation area
6.	Under utilized/degraded notified forest land
7.	Degraded pastures/ grazing land
8.	Degraded land under plantation crops
9.	Sands- Desertic/ coastal
10.	Mining/industrial wastelands
11.	Barren Rocky/Stony waste/Sheet rock area
12.	Steep sloping area
13.	Snow covered and/or glacial area

Remote Sensing and GIS Analysis for Wasteland Studies

There have been growing awareness regarding wasteland development and a number of studies have been conducted to exam the problems and reclamation of wastelands. It has been especially true in Indian panaroma. Previously, the recommendations to reclaim wastelands were on the basis of local experience and with limited local field observations. At this juncture, the use of remote sensing and geographic information system were well appreciated since they could be applied to approach the problem at the regional as well as at the local level. They also aid in analyzing various terrain conditions to plan scientific reclamation and management practices. A few of such studies are discussed here.

Singh (1969) and Jasbir Singh (1971) have studied the spatio-temporal patterns of wastelands using conventional surveys and existing collateral data. Singh (1971) suggested conducting micro-level studies to understand the extent and problems pertaining to wastelands. Later, the use of remotely sensed data played an important role in mapping the wasteland units. Sen (1972) used aerial photographs combining with intense field verifications depicted the advantage of using remote sensing data. Montapa (1972) and Sankaranarayan and Singh (1979) were able to recognize and classify wasteland units using aerial photographs. Further, Gautam et al., (1986) demonstrated the use of satellite data in conjunction with collateral data for wasteland development programmes and evolving management strategies. The data derived from satellite were useful in gathering information on wastelands, which helped in generating a wasteland database by Government Agencies (Oberai and Singh, 1986). In another study carried out by Gupta et al., (1996), the application of remote sensing in inventorying degraded lands is emphasised by visually delineating various land units such as agriculture, forest, open scrub, erosional surfaces, kankars and saline areas using IRS- 1A LISS-II data.

Mruthanjaya et al., (1992) explains the applicability of GIS by spatially integrating various themes for resources planning. The feasibility of using GIS for assessing land's potential for wasteland development at Block level on 1:50,000 scale has been demonstrated by Nagarathinam and Hamsanandan (1993) in a pilot study. Geetman et al., (1995) related the natural planning and management especially for wasteland reclamation purpose. A farm level development planning, with the aid of GIS techniques, was explained by Sivasamy et al., (1997). The study was conducted in a small village (Vadakkur village) in Tamil Nadu, India emphasized the utility of GIS in micro-level analysis, planning and land management.

Reclamation of Wasteland Studies

Sankaranarayan (1988) suggested agro-forestry as means of reclaiming wastelands, which would further protect the land from degradation. Wastelands in terms of animal husbandary may provide sustenance to many animals by fodder cultivation. This condition holds true in Rajasthan, where a meager unfit land may be sufficient to feed domestic livestock (Ghosh and Mittal, 1988).

The success of wasteland reclamation is based on micro-level planning with a clear understanding of topography, soil type, erosion, rainfall distribution and socio-economic conditions of local community (Hegde, 1993). Bhavani et al., (1996) concluded that the nutrient value of land and socio-economic aspects of farmer were the two major problems of wastelands in Tamil Nadu. This stressed the importance of economic conditions of farmers as an important input in the wasteland analysis. Inter-relationship between agriculture, forest and wastelands was explained by Shah (1995). A few of wasteland reclamation activities like growing salt-tolerant species were suggested by Rai (1999).

Pisharoty (1987) and Singh and Mittal (1993) stressed on the importance of water and watershed management to wasteland development activities. They stated the utility of considering rainfall distribution, water level variations, soil condition and socio-economic condition of local community as factors in wasteland developmental activities. Nagarathinam (1997) explained the usefulness of remote sensing and GIS based approach to assess land potential for wasteland development at village level in a study conducted around Kalluthu village, Usilampatti Block of Tamil Nadu, India. Vijay and Pradeep Kumar (2001) integrated various thematic layers such as lithology, landforms, soil, groundwater level, rainfall and slope to understand their influences on wastelands and also explained the utility of remote sensing and GIS techniques for reclaiming them.

Objectives

Keeping the above mentioned scenario in mind, an attempt has been made in the present study to integrate remote sensing and GIS methods with the following objectives:

- to identify , classify and quantify the wastelands using remote sensing technique,
- to evaluate the land quality and water resource characteristics of the study area for assessing the resource potentials using Geographic Information System (GIS),
- to suggest the site specific action plans for cadastral-level units of for reclamation of wastelands, based on land potentials and socio-economic factors and
- to demonstrate the usefulness of the science and technology to the local farmers

Methodology

Integrating remote sensing and collateral data using GIS techniques for resources identification and wasteland development is the main aim of the study (Figure 1). This approach is used to understand various terrain settings such as geology, geomorphology, soil and land use patterns of the study area, the Pulimankombai micro-watershed. Collateral data include rainfall, water level, water quality, socio-economic details, soil details, relief, drainage and transportation network of both spatial and temporal information on the study area. Data on water level, water quality and rainfall are collected from various Government agencies (Public Works Department, Agricultural Department and Revenue Department of Government of Tamil Nadu) and also from field surveys. The data were collected for a period of ten years from 1991 to 2000. Isopleth were generated for all the above collateral data with appropriate interval and time gap based on season. For example, water level data of summer, winter, southwest monsoon, and northeast monsoon were selected to analyse the seasonal fluctuations over a period of time. Similarly, isopleth were generated for water quality parameters and rainfall. These maps were transformed into a GIS platform and used in the analysis by giving proper weightage along with other thematic maps such as geology, geomorphology, soil, landuse and wastelands. The landuse maps were also space and time specific and derived from interpreting satellite data and field verifications. Wasteland units form a part of landuse maps and were separately extracted for analysis in variation of aerial extent of wasteland units over a period of time. Integrating all these informations, analyses were carried out and results were brought out.

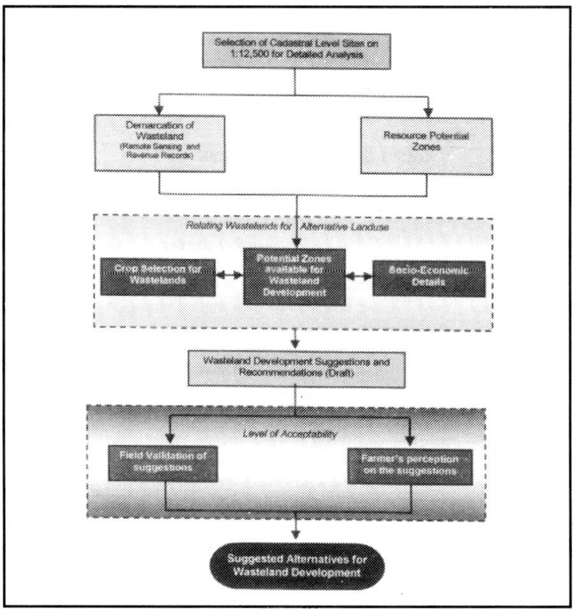

Fig. 1. Methodology for Arriving Alternative Measure for Wasteland Development in the Micro Watershed

Selection of Study Unit

Analysis using remote sensing and GIS technique for wasteland reclamation is carried out in the selected cadastral micro-watershed. This approach for wasteland is advocated by Kulandaiswamy (1997) in High Level Committee on Development of Wastelands, watershed development and implementation of poverty alleviation programmes and is currently being adopted by the Tamil Nadu Watershed Development Agency (TAWDEV, 2001). A smaller Pulimankombai mini-watershed of Nagalar sub-watershed, Vaigai Basin has been selected for the present GIS analysis. Pulimankombai mini-watershed is located in the northeastern part of the study area. It occupies an area of 4212.36 ha. It is surrounded by Andipatti RF Northern hills on all its three sides and bound by Vaigai Main River in the north. It has a forest area of 1614.12 ha. and agricultural area (non-forest area)of 2598.24 ha., which are confined to the south and north respectively. The mini-watershed truly represents the entire study area as it has a varied landform characteristic and also all the wasteland units such as permanent fallow, salt affected, land with or without scrub and degraded forest.

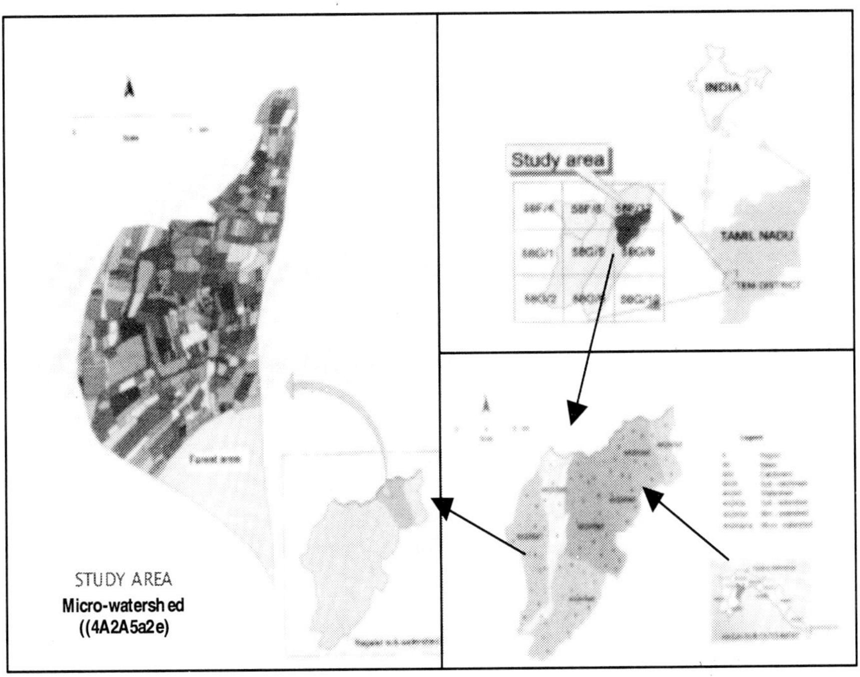

Fig. 2. Location of selected micro - watershed (4A2A5a2e) of Kilimankombai mini-watershed

Most of the drainage in Pulimankombai mini-watershed run from south to north. Sources of these drainage are mostly from adjoining hills (Andipatti RF Northern hill slope) and joins Vaigai main river in the north. In this way, it is ideal since it has source as well as destination within the mini- watershed. The agricultural activity in the mini-watershed mainly depends upon the drainage from first order to third order streams and there is no major surface storage.

There are seven micro watersheds within this mini-watershed from which one micro watershed has been selected for the cadastral level analysis. For detailed investigation and analysis larger scale information are required and hence, information obtained from large-scale remote sensing data is combined with field data.

The details such as field sub-division and ownership information are obtained from the Village Administrative Officer (VAO) of the selected village. Large-scale field sub-division map from Field Measurement Book (FMB) sketch on 1:8,000 scale and the respective field details from the Adangal register available with the VAO are taken as GIS input after transforming the scale to 1:12,500. Thus, a base map showing the field subdivision boundaries was completed in order to plot such attribute data as land ownership and other related socio-economic aspects on 1:12,500 scale.

Socio-economic Condition in the Selected Micro-watershed

A revenue village, Pulimankombai, partly coincides with the boundary of the selected micro-watershed. Hence, the village level details pertaining to people, occupation and their economic condition reflect the utilization of resources and other developmental activities within the micro-watershed. The population details were collected from the published report by the Census Operations, Govt. of India. Other socio-economic details were collected from the National Informatic Centre (NIC) of Theni District.

Population details collected for the past four decades (from 1961 to 1991) show large increase in population from mere 1378 to 3027 indicating that the growth is more than twice the total of population during the observed periods. The population density also increases from 80 to 176 per sq. km during the period. At the same time, number of cultivators in the village shows only a marginal increase (from 690 to 778) with many of the villagers depend upon agriculture for their sswo cadastral sites and are discussed in the following section.

Soil Mapping through Detailed Soil Survey (Dss)

In the cadastral level analysis of wasteland reclamation, a detailed soil survey is a pre-requisite (Sharma et al., 2001). To derive a clear understanding on the soil condition and their subsequent suitability for reclamation, soil survey has been conducted in the micro-watershed over a period of 21 days. Geology, geomorphology and slope are considered as base line to delineate individual morphic units (Elbersen et al., 1988; Zinck and Valenzuela, 1990; and Ramamurthy et al., 2001) and points for collecting soil samples are identified. Using the methods of clino-sequence and photomorphic units 36 mini-pits and 12 master pits were selected and examined, and a total of 48 pit-samples were collected according to their horizon characteristics and analysed with the help rendered from the Agricultural Officers of the Soil Survey and Landuse Organisation (SS & LUO), Government of Tamil Nadu, Thanjavur. The morphological and analytical soil characteristics derived from the samples helped to identify the soil taxonomy using USDA soil survey manual as reference. But for the local level analysis, series level information is required, which is again derived from soil taxonomy. This way, nine soil series are identified in the cadastral site and mapped on 1: 12,500 scale (Figure 3). From the soil series, land capability class, land irrigability class and crop suitability are determined.

Soil Series of the Micro-watershed

Pulimankombai series has very deep, brown in color, sandy clay loam surface and sandy clay subsurface texture, with 7.1 pH and moderate to slow permeability. It occurs in the flood plain on charnockite landscape of below 1 per cent slopes. It occupies 13 per cent of the study area. Maravapatti series is very deep, yellowish red to reddish brown in color, sandy loam surface and gravelly sand clay subsurface texture, pH 8.4 and moderate to slow permeability. This soil is derived from buried pediment medium on garnetiferous sillimanite gneiss landscape of 1-5 per cent slopes. It occupies 32.2 per cent of the study area. Rajagopalanpatti series is shallow, yellowish red to reddish brown in color, gravelly sandy loam surface and gravelly sandy clay loam subsurface texture, pH 6.1 and have rapid permeability. This soil is developed over buried pediment shallow on gneissic granite landscape of 1-5 per cent slopes. It occupies 20.3 per cent of the study area. Pudur soil series is very shallow, yellowish red in color, gravelly sandy clay loam surface texture, pH 7.5 and have rapid permeability .This soil is derived from pediment on charnockite landscape of 5-10 per cent slopes and it occupies 11 per cent of the study area. Valuttu soil series has shallow, dark reddish brown in colour, sandy loam surface texture, pH 6.2, and have very rapid permeability. This soil is derived from structural hill on charnockite landscape of above 10-15 per cent slopes and it occupies 23.5 per cent and of the study area.

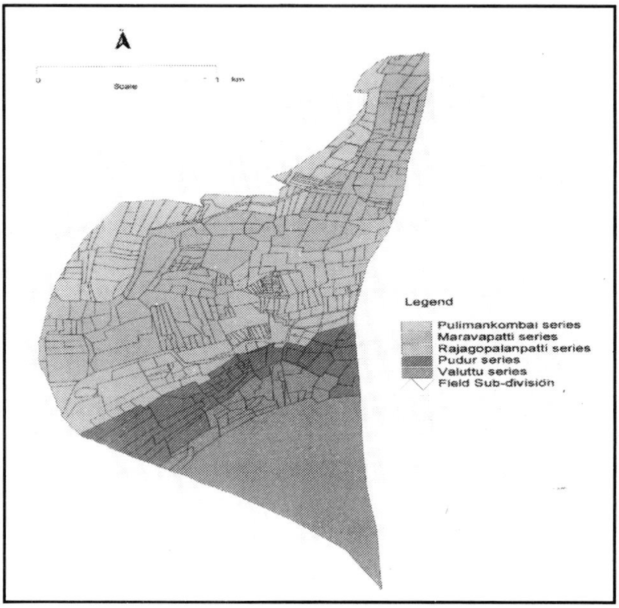

Fig. 3. Soil through detailed soil survay

Land Capability Classes (LCC)

Land capability classification is an interpretative grouping of soils mainly based on their inherent soil characteristics, external land features and environmental factors that limit the use of land for arable agriculture. The land capability classes place the soils into general order of suitability or otherwise for agriculture, forestry, grassland or other uses for sustained production.

LCC of Maravapatti soil series show good cultivable land with moderate limitations. These are calcareous soils and show some soil problem that requires conservation in irrigation methods and improved drainage. Selection of suitable crops and careful cultivation practices may improve the condition. Pulimankombai soil series falls under LCC showing moderately good cultivable land. This is similar to Maravapatti series and show soil limitation requiring improved drainage and irrigation methods. Pudur soil series falls under moderately good cultivable land. But these soils show severe erosion and leaching away of mineral constituents. Soil reclamation requires erosion control measures. Agriculture may be practiced after adopting soil conservation and carefully selecting the crops. Valuttu soil series are fairly cultivable lands and show severe limitation. Erosion of surface material is predominant leaving gravels and stones. Requires severe soil conservation practices. Cropping is possible with some limitations. LCC of the soil series in the cadastral area (Figure 4) implies Maravapatti, Pulimankombai, and Pudur soil series could be used for agriculture related wasteland reclamation activities.

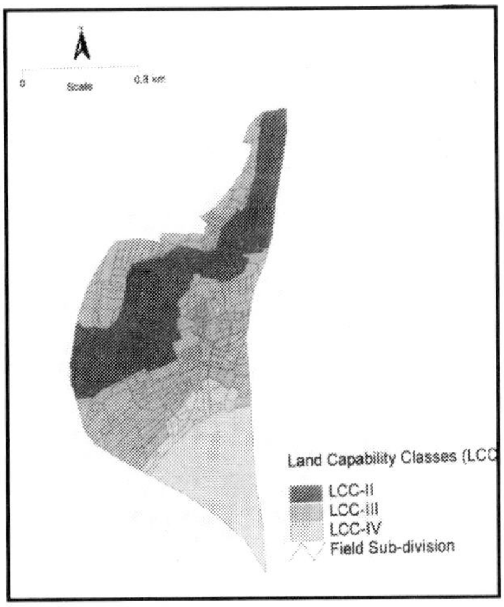

Fig. 4. Land capability classes (LCC)

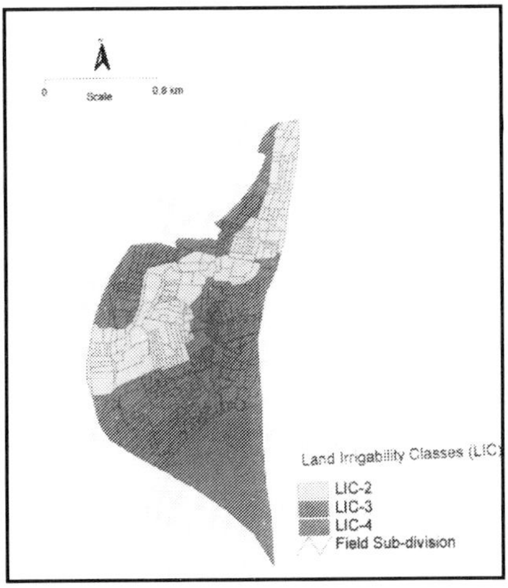

Fig. 5. Land Irrigability classes (LCC)

Land Irrigability Classes (LIC)

The irrigability classification is primarily aimed at predicting the behavior of soils under the greatly altered water regime brought about by introduction of irrigation. The interpretation governed by two sets of factors namely 1. The inherent soil characteristics and 2. Other external features and non soil related factors such as slope, topography, drainage development etc. The land irrigability classes are worked out according to irrigability depending on physical land features, socio-economic conditions, quality of water and drainage requirement for sustained use under irrigation (Figure 5).

Maravapatti soil series show moderate and severe limitations of land irrigability under sustained agricultural practice. The water holding capacity of the soil series requires the adoption of such measures as terracing, vegetative bunding and mulching. Pulimankombai soil series show severe limitations for sustained use under itrigation. The conservation measures of this series are similar to Maravapatti. Pudur soil series show severe land irrigability limitations. Topography operates as a major constraint in this soil series. Valuttu soil series also have very severe soil limitation and topography plays a major role in soil limitation. Soil may be conserved after adopting severe conservation measures and irrigation practices. This soil series may be best suitable for agro forestry.

Ranking of the Soil Series

Soil capability and soil irrigability depend upon the soil characteristics. Ranking based on such soil characteristic help in arriving four soil suitability zones viz., Soil group A, Soil group B, Soil group C and Soil group D (Table 2). These zones reveal the degree of soil suitability for wasteland reclamation activities in the cadastral villages. Ranking of soil groups indicate that the higher degree of soil suitability starts with soil group A and it progressively decreases with soil group B, C and D. A brief description on these soil groups is given below.

Table 2. Ranking of soil Series according to their charcteristics

S.No	Soil Series	Land Capability Class	Land Irrigability Class	Rank
1.	Pulimankombai	IIIs	3s	B
2.	Maravapatti	IIs	2s	A
3.	Rajagopalanpatti	IIIst	3s	B
4.	Pudur	IIIst	3se	B
5.	Valuttu	Ivse	4t	C
6.	Subbulapuram	IIIst	3se	B
7.	Vairamuttu	Ves	5t	C
8.	Kalugumalai	Vies	5t	D
9.	Nagamalai	Ves	4t	C

Soil group' A' comprises of Maravapatti series. Though this series shows higher degree of suitability, its association with RPZ2 having permanent fallow and salt-affected land, indicates some constraint in wasteland reclamation. Soil group 'B' comprises of Pulimankombai, Rajagopalanpatti, Pudur and Subbulapuram soil series. Among the soil series in this group, Pulimankombai shows its association with RPZ I. This may be due to intense agricultural practice in this zone without adhering to any effective conservation measures, causing the soil loose its soil richness. These causes increase in salinity in the soil as well as decrease in land productivity leading to the condition of permanent fallow. The rest of the soil series in this group is associated with RPZ2, which have permanent fallow and land with scrub. Soil group 'C' consists of Valuttu, Vairamuttu and Nagamalai soil series. In this group, Vairamuttu soil series fall in RPZ3 and requires intense conservation measures before adopting reclamation practices. It is seen with scrub and degraded forest. Valuttu and Nagamalai soil series are associated with RPZ4 and are observed with stony waste, scrub land and degraded forest. Soil group 'D', which is mostly suitable for forest related developments, comprises of Kalugumalai series and is associated with RPZ4 having degraded forest.

Ranking of soil suitability indicate a closer association with the presence of wasteland categories in the study area. For example, soil group A have permanent fallow and salt-affected land where as the lesser degree soil group are mostly associated with scrub land, degraded forest and stony waste. Presence of permanent fallow and salt-affected land in soil groups A and B again indicate the role of economic constraints in this region apart from certain physical constraints. The above observations and discussion on soil series show degree of limitations imposed on wasteland reclamation activities. Moreover, it requires additional information on wasteland units at cadastral level for reclamation. Hence, larger remote sensing data and conventional data are used to extract field level information on wasteland is discussed in the following sections.

Extraction of Wasteland Categories using Conventional Data and Satellite Data

Wasteland map has been prepared at village level using IRS-1D satellite merged data (PAN + LISS-III). Map showing field sub-divisions are compared with the remote sensing derived wasteland map. During such analysis, a field sub-division occupied by more than 75 per cent of a particular wasteland unit (for example permanent fallow or salt affected) is categorized as the same wasteland unit. Similarly, a permanent fallow land occupying 75 per cent of a field sub-division may be identified as a permanent fallow. This method is adopted for all the land parcels and in this way, field sub-divisions are categorized to show various wasteland units. It also helps in updating wasteland information. Field sub-division wise wasteland information that is available in the official record (Adangal Register) were collected and mapped on 1:8,000 scale.

The final map thus generated shows five wasteland classes while combining 'Adangal ' register (revenue record) classes and satellite data derived classes. There were only two wasteland categories such as fallow and stony waste as per revenue classification, but when remote sensing data classes are superimposed, the following five classes were obtained, such as 1. Permanent fallow, 2. Salt-affected land, 3. land with/without scrub, 4. Degraded forest and 5. Stony waste. These maps reveal more wasteland categories than the map of revenue classes, alone which are not sufficient for detailed studies and hence form as basis for village level wasteland analysis. The respective attribute data is also updated so that each land parcel may have a particular landuse unit with respect to wasteland map derived from remote sensing data. Based on the updated statistics the study area is having an areal extent of 144.99 ha. (66.14 per cent)

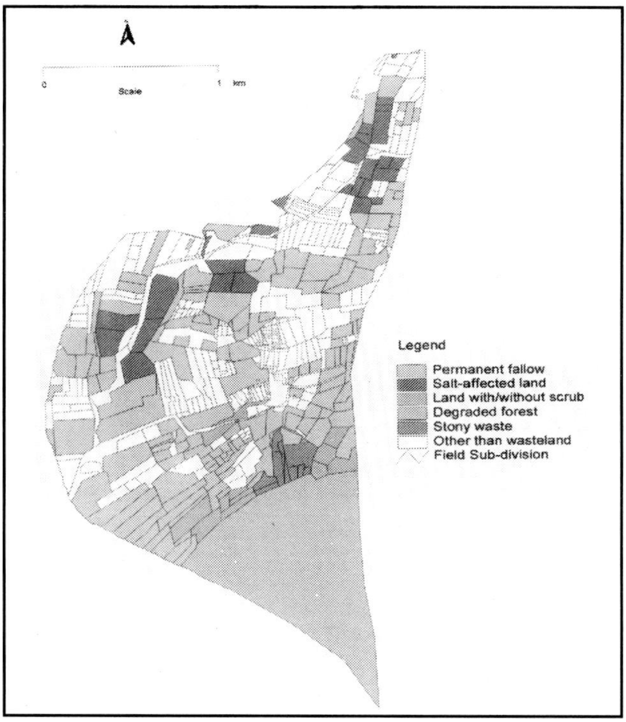

Fig. 6. Wasteland derived from Stallize Images and Revnue Records

Land Potential Zones (LPZ)

To understand the potentiality of land resources in the study area, GIS overlay analysis is carried out. Various thematic layers such as geology, geomorphology, soil and slope are considered for the analysis. These layers are the indicators of physical conditions of a terrain or land. Geology of the terrain is considered as a basic unit and accordingly, the significance of slope and landform units of the particular lithological environment are related. This relationship forms the basis for zonation of land potential. For example, a landform of a particular slope may develop certain soil type under particular lithologic and climatic conditions, which may be used for some specific landuse (Majid Ghiassi Far, (2001). Based on this principle, categories in each thematic layer are given proper weightage according to their landuse capability significance and are added to derive the Land Potential Zone map.

Based on the above analysis, of LPZ are recognized five zones viz., LPZ-1, LPZ-2, LPZ-3, LPZ-4, and LPZ-5. The Excellent degree of land potential is shown as LPZ-1; very good degree of land potential as LPZ-2; good degree of land potential as LPZ-3; moderate and poor as LPZ-4 and LPZ-5 respectively.

Water Potential Zones (WPZ)

The GIS overlay technique is carried out to generate zones showing different water potentials from WPZ-1 (excellent), through WPZ-2 (very good), WPZ-3 (good), and WPZ-4 (moderate) to WPZ-5 (poor). Apart from considering physical parameters such as landforms, slope, and hydrological soil group, information on groundwater details (well yield and resistivity) and drainage density are also considered in grouping WPZ. Based on the significance of each parameter, categories in each thematic layer are given proper weightage and added to derive the Water Potential Zone map.

Resource Potential Zones (RPZ)

Resource potential of the study area includes both land and water resources. Analysis and discussion of these two resources in the previous sections helped to understand their individual relationships to landuse practices of the study area. In practice, it is essential to understand their cumulative effects and their influence on land utilization (Douglas, 1994).

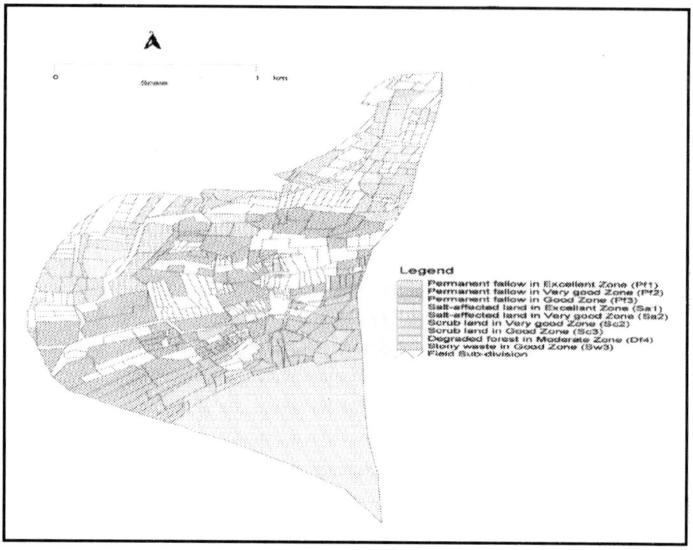

Fig. 6. Wasteland derived from Stallize Images and Revnue Records

The resources potential of the study are is derived from the GIS overlay analysis of both LPZ and WPZ and the resultant maps show five zones of Resource Potential viz., RPZ-1 (excellent), RPZ-2 (very good), RPZ-3 (good), RPZ-4 (moderate) and RPZ-5 (poor). Among these RPZ-1 indicates zone having excellent land potential and excellent water potential. The RPZ-2 as very good land potential and very good water potential and so on.

The Resource Potential Zone of excellent category has some areas in Permanent fallow of wasteland map, and some in salt-affected land domain (Table 3). These are termed as Pf1 and Sal, and they occupy an area of 11.02 ha., 9.76 ha. Very Good Zone of RPZ occurs over Permanent fallow (Pf2), Salt-affected land (Sa2) and Land with / without scrub (Sc2) occupying an area of 104.48ha., 5.35 ha. and 59.2 ha. In Good Zone of RPZ, Permanent fallow (Pf3), Land with / without scrub (Sc3) and Stony waste (Sw3) are formed to occur, occupying an area of 12.13 ha., 74.81 ha., and 323.17 ha. Moderate Zone of RPZ occurs over Permanent fallow (Pf4), Degraded forest (Df4) and Stony waste (Sw4) category of wasteland, occupying an area of 2.88 ha., 439.87 ha. and 4.72 ha. This classification brings out the Potential Zones Available for Wasteland Development (Figure 7), also help in adopting suitable measures for improving the productivity of wastelands.

Table 3. Wasteland Categories in Resources Potential Zones

RPZ	Area	Pf	Sa	Sc	Df	Sw	Total
Zone-1	43.15	11.02 (7.6%)	9.76 (6.7%)	–	–	–	20.78 (14.3%)
Zone-2	109.63 (38.6%)	55.98 (3.6%)	5.35 (0.5%)	0.81	–	–	62.14 (42.9%)
Zone-3	25.41	12.13 (8.4%)	–	6.26 (4.3%)	–	3.47 (2.4%)	21.86 (15.1%)
Zone-4	41.00	–	–	–	40.21 (27.7%)	–	40.21 (27.7%)
TOTAL	219.19 (54.6%)	79.13 (10.4%)	15.11 (4.9%)	7.07 (27.7%)	40.21 (2.4%)	3.47 (66.14%)	144.99

Suggestions for Wasteland Development Based on GIS Analysis

Six major landuse categories or types are recommended for effective reclamation of wasteland in the study area, viz., 1. improved agricultural activity, 2. agro-horticulture, 3. agro-forestry, 4. horticulture, 5. fodder / fuel wood plantation and 6. forest gap plantation.

According to the above categories, 'improved agriculture activities' include intense agriculture activities with already available soil and water resources or by adopting soil conservation measures. Cropping pattern may include paddy, sugarcane, cotton, millets, ragi and cholam. The term 'agro-horticulture' includes mixed agricultural practices with horticultural species. This may include growing mangoes or citrus fruit trees along with cultivation of dry crops. 'Agro-forestry' includes growing of trees (teak) along with cultivation of millets and pulses. This activity is most preferred for wastelands development in areas adjoining croplands (Sankaranarayan, 1988). The term 'horticulture' mainly indicates growing of fruit trees such as pomegranate, guava, mango, and citrus fruits apart from vegetables, flowers and herbs.

The above land use categories may be recommended to reclaim various wasteland units with either 'minimal effort' or 'maximum effort'. Here, the minimal efforts mainly indicate physical constraints such as soil suitability and availability of water for irrigation. Many of the farmers in this area depend upon either the monsoon for irrigation or tanks. This may be fruitful during rainy season but during summer many of the parcels of land are left uncultivated due to lack of enough surface storages for irrigation. Thus, many of the farmers adhere to traditional cropping pattern, cultivating only a particular type of crop, instead of crop rotation causing the land loose its soil nutrients. This may be remedied by crop rotation, soil conservation and use of modern farming techniques, which requires investment by the farmers, which comes under the purview of 'maximum effort'. The maximum effort also includes digging of borewells, drip irrigation, crop management (nurseries, manures and pesticides), market potential, trade and transport, and investment for long-term produces such as agro-forestry and agro-horticulture.

Based on the above descriptions and combining with the information derived from GIS techniques, suggestions for reclamation of wastelands in both the cadastral sites have been discussed in the following sections.

Suggestion for Micro-watershed

This micro-watershed is farming oriented with 178.98 ha. of land under cultivation and 40.21 ha. under forest. Suggestions for wasteland development based on the available resources in the area, soil condition, slope condition and groundwater potential are arrived and is shown in Table 4. Based on the analysis of these two maps only the above two options, viz., 'minimal effort' and 'maximum effort' are suggested (Figure – 8&9 : Table – 5). The wasteland units and suggested reclamation activities are discussed in the following sections.

Permanent Fallow: Permanent fallow is found to occur in all the resources potential zones from zone 1 to zone 3) except in zone 4. Thus, in the Pf1, excellent zone has Pulimankombai and Maravapatti soils series. Slope is flat to gently undulating with very good groundwater potential. Soil type and availability of groundwater indicates that Pf1 could be easily reclaimed. The land potential is as good as the cropland; it remains as permanent fallow due to some socio- economic constraints. During field investigation it was found out that lack of fund was the main reason. Hence, this category of wasteland can be developed into productivity by bringing it under cultivation. The area under this category is 11.02 ha. Permanent fallow falling in zone 2 (Pf2) has both Maravapatti and Rajagopalanpatti soil series with a slope percentage ranging from 1 to 5. Groundwater

Table 4. Suggestions for the Wastelands Development

Wasteland category	Wasteland unit symbol	RPZ	Soil Series	Slope in percentage	Water Resource	Suggestions Minimum effort	Maximum effort
Permanent Fallow	Pf1	Excellent	Pulimankombai Maravapatti	0-1	Very good	Imrpvoed agricultural activities	Imrpoved agricultural activities
	Pf2	Very good	Maravapatti Rajagopalanpatti	1-5	Very good-Good	Agro-horticulture	Improved agricultural activities
	Pf3	Good	Pudar	5-10	Good-Moderate	Agro-forestry	Agro-horticultural
Salt-affected land	Sal	Excellent	Pulimankombai	0-1	Very good	Agro-forestry	Improved agricultural activities
	Sa2	Very good	Maravapatti Rajagopalanpatti	1-5	Very good-Good	Agro-forestry	Agro-horticultural
Scrub land	Sc2	Very good	Maravapatti Rajagopalanpatti	1-5	Very good-Good	Agro-forestry	Agro-horticultural
	Sc3	Good	Pudur	5-10	Good-Moderate	Fodder/ Fuelwood Plantation	Medicinal plants
Stony waste	Sw3	Moderate	Valuttu	10-15	Moderate-Poor	Fodder/ Fuelwood plantation	Medicinal plants
Degraded forest	Df4	Moderate	Valuttu	10-15	Moderate-Poor	Forest gap plantation	Forest gap plantation

resource is very good to good. The wastelands falling in this zone may be improved by practicing agro-horticulture method effort and could also be used for improved agricultural activities after adopting adequate conservation measures. The area under this category is 55.98 ha. (38.6 per cent). Permanent fallow falling in zone 3 (Pf3) has Pudur soil series and moderate slope having moderate groundwater resource. This wasteland unit is suitable for agro-forestry and could be reclaimed for agro-horticulture after applying meticulous conservation measures. Area under this category is 12. 13 ha., (8.4 per cent). Hence, restoration of cultivation is the chief mode of reclamation suitable for this zone.

Salt-affected Land: Salt-affected land is a wasteland category, which has lost its productivity only temporarily. If proper measure of reclamation is applied, the condition may be reversed. Salt-affected land is mostly confined to resource potential zone 1 and zone 2. The cause for salinity of the soil may be due to lithology, calcareous in nature and excessive surface erosion.

Salt-affected land that is observed in excellent resource potential zone (Sal) has predominantly Pulimankombai soil. The terrain is flat with gentle undulations and the groundwater resource is very good.

Table 5. Alternative for Wastelands Development (Table contains samples extracted from the derived final output)

Field Sub-division no.	Area	Owners	Wasteland unit	Suggestions Minimum Effort	Maximum Effort
22002A	0.83	Lingusami. P.	Pf1	Improved Agricultural Activities with Shallow dug well irrigation (Paddy, Millets, Vegetables and Floriculture)	Improved Agriculture Activities with deep bore well irrigation (Banana, Sugarcane, Paddy and Betelnut)
24301	1.74	Periyabommi Naicker. P	Pf2	Agro-horticulture with Shallow dug well irrigation (Guava, Pomegranate, Papaya Mango, Citrus and Murunga along with Groundnuts, Minor millets and Pulses)	Improved Agriculture Activities with deep bore well irrigation (Paddy, Coconut, Banana and Sugarcane)
34305	0.88	Rengammal K.	Pf3	Agro-forestry with Shallow dug well irrigation (Teek, Neem, Vagai along with Groundnuts, Minor millets and Pulses)	Agro-horticulture with deep bore well irrigation (Mango, Citrus, Guava, Pomegranate and Mununga along with Groundnuts, Minor millets)
22702	0.93	Sundara Mahalingam R.	Sal	Agro-forestry with salt tolerant crops (Casunina, Euclyptus, Neem along with ragi, Chilies, Cotton and Pulses)	Improved Agriculture Activities with soil conservation measures (Paddy, Ragi, Cotton, Tobacco and Chilly)
25902	0.81	Vegattammal S	Sa2	Agro-forestry with salt tolerant crops (Casunina, Euclyptus along with ragi. Chilly, Cotton and Pulses)	Agro-horticulture with soil conservation measures (Guava, Pomegranate and Murunga along with Groundnuts, Minor millets)
33402	.051	Rahmathulla A	Sc2	Agro-forestry with Shallow dug well irrigation (Teak, Neem, Vagai along with Groundnuts, Minor millets and Pulses)	Agro-horticulture with deep bore well irrigation (Guava, Pomegranate and Murunga along with Groundnuts, Minor millets)
34901	1.13	Government	Sc3	Fodder/Fuel wood plantation (Grasses Casurina, Odai, Kokkimul, Mansevivel and Prosopies)	Medicinal Plants with deep bore well irrigation
34702	0.53	Samuvel K	Sw3	Fodder/Fuel wood plantation (Grasses, Casurina, Odai, Kokkimul, Mansevivel and Prospies)	Medicinal Plants with deep bore well irrigation
		Government	Df4	Forest gap plantation (Odai, Acha, Moongil, Arappu and Perumaram)	Forest gap plantation with soil conservation measures (Tamarind, Teak, Moongil, Acha)

Pf : Permanent Fallow
Sa : Salt-affect land
Sc : Scrub land
Df : Degraded forest
Sw : Stony waste

The land may be used for agro-forestry and could be utilized for improved agricultural activities after adopting soil conservation measures. Area under this category is 9.76 ha. (6.7 per cent).

Similarly, Sa2 is observed to be associated with Maravapatti and Rajagopalanpatti soils series with good groundwater potential. The land may be used for agro-forestry with soil improvement but could also be utilized for agro-horticulture practices when sufficient conservation measures are applied. Soil conservation is the main problem in these units and when appropriate measures are adopted, the land may be well used for growing fruit trees. Area under this category is 5.35 ha.,(3.6%).

Land with / without Scrub: Land with / without scrub is associated with resources potential zone 2 (very good) and zone 3 (good) and the terms Sc2 and Sc3 are used to describe them respectively. Sc2 is observed with Maravapatti and Rajagopalanpatti soils series

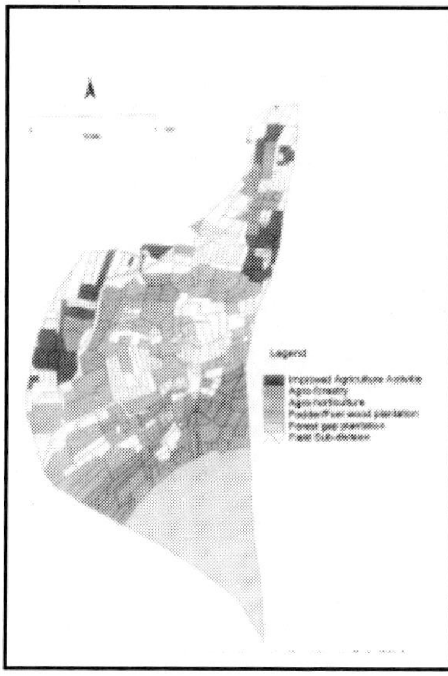

Fig.8. Suggestion for Wasteland Develpment with Minimum Effort

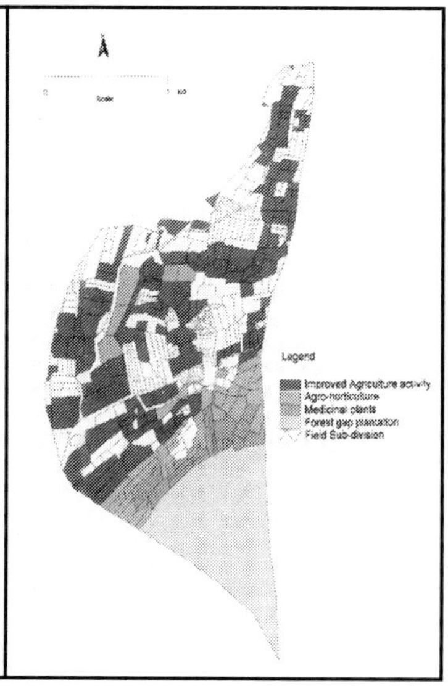

Fig 9. Suggestion lal Wasteland Develpment with Maximum Effort

with moderate sloping terrain. Groundwater potential is very good. Land units may be used for agro-forestry. It may also be utilized for growing fruit trees, vegetables and flowers with some limitations and after adopting soil conservation measures. Area under this category is 0.81 ha. (0.5 per cent).

Land with / without scrub land falling with good resources potential (Sc3) is mainly comprised of Pudur soil series with moderate to inclined slope condition. Only moderate potential of groundwater is observed in this region and the land units could be used for growing fodder and fuel wood plantation trees. With sufficient measures the land may also be used for agro-forestry practice. Area under this category is 6.26 ha. (4.2 per cent).

Degraded Forest: Degraded forests are mainly confined to the moderate resources potential zone (Df4) with steep slope and immature soil series, Valuttu. They may be used to grow forest related trees and plants and utilized for forest gap plantation varieties. Area under this category is 40.21 ha. (27.7 per cent).

Stony Waste: Stony waste falling in zone 3 (Sw3) observed with Valuttu soil series and steep slope. Ground water potential is also ranging from good to moderate. This is not suitable for any meaningful land use practice and fodder and fuel wood plantation trees and plants may be grown. Area under this category is 3.47 ha. and a percentage of 2.4 to the total wasteland area. All the above wasteland units and suggested reclamation activities based on the characteristics of soil series, condition of slope, and groundwater potential are given in Table 4. As a further step, while giving the appropriate suggestions, Field Survey numbers, area of the fields, ownership details, and suitable crops and species are added and tabulated (Table 5) which would serve as a ready reckoner for planners.

Field Validation of Suggestions

The transformation of details of available resources from smaller scale to cadastral level shows a slight degree of discrepancy, which was corrected by means of field verification. Analysis combining GIS derived available resources with the cadastral level wasteland units as derived from remote sensing data are also checked in the field. Detailed field investigation was carried out to correlate the GIS based suggestions for reclamation of the wasteland parcels. Categories of permanent fallow under different resources zones are verified in the field. The areas adjacent to Pfl, which have similar, soil series are seen with agricultural activities, indicate possibility of reclamation with improved agricultural activities. In some land units, certain soil conservation measures are necessary. Land parcels around Pf2 have mostly dry crops and fruit tress and in some places even paddy. These again corroborate well with the suggestions derived from the above analyses.

Similarly, many of the salt -affected land parcels in the cadastral sites are used for agro-horticulture, agro-forestry and even crop cultivation after adopting adequate conservation measures. This again corroborated with the suggestions that wasteland units of this category may be utilized for growing fruit trees along with dry crops or vegetables and flowers along with millets. Scrublands are mostly observed along the foothills and are closely associated with thorny trees and grass, which is used as grazing land by the local community. It is suggested that, these land parcels may be utilized for growing fuel wood plantation and fodder. A smaller patch in the central part, which is surrounded by agricultural land units indicate the possibility of reclamation of this particular unit for growing vegetables, flowers and citrus trees after intensive conservation measures. In the field, the areas adjoining stony waste are observed with growth of thorny trees and herbal plants, which again indicate the possibility of an optimal utilization of land parcels by growing fodder, fuel wood plantation and herbal plants.

During field verification, it was seen that almost all the wasteland units show similar relationship with the GIS derived resources map as shown in Table 4 and Table 5. From the field observation, it is also noted that many of the wasteland units are the products of negligence of farmers owing to physical and economic constraints, especially the presence of permanent fallow along the river course. The difficult part of field validation lies in assessing the cause of wasteland since many of them are due to socio-economic condition of the local community. Physical parameters such as soil and water, do not pose a major problem, since land use pattern adjacent to wasteland units show some sort of productive land uses. For example, most of the areas adjoining permanent fallow show good agricultural activities. From the field visits in the Pulimankombai mini-watershed, food crops (such as paddy, ragi and millets), cash crops (banana, sugarcane and groundnut) horticultural varieties (mango, guava, citrus, papaya, sunflower and vegetables), inter-cropping (coconut and silk cotton with dry crops) and presence of teak and neem trees are observed. Few patches of cashew and eucalyptus plantation are also observed. The existing land use pattern indicated the possibility of utilizing the wasteland units for anyone of the above-mentioned landuse. There is every possibility of motivating the local farmers to adopt reclamation activities with minimal effort, since the land has the necessary potential for carrying out the above suggestion. This necessitated the need to interact with the local people in the cadastral sites, during the verification process. For this purpose counseling was held with the farmers who have been selected from the ownership details pertaining to the problem lands. The villagers are interviewed about adopting suggested reclamation activities in their lands, which is discussed in the following section.

Farmer's Perception

Through councelling and interaction sessions, farmer's perception on the suggested reclamation measures has been assessed. The ownership information and suggested reclamation activities of land parcels are extracted from the attribute data of the GIS derived maps. This information, thus, forms the guidelines to identify the owners for interaction. For example, a land unit that has been under Pf1 belonging to a particular farmer has been met in person and discussed about the suggestion for his acceptability. This approach was carried out for land parcels that have been selected randomly across the study area and the concerned farmers are interviewed. Through the interviews, marginal farmers, small farmers, medium and large farmers are segregated based on their land holding capacity. Many of the marginal and small farmers have land in the same village whereas large farmers posses land in other villages too.

Besides, Agricultural Extension Officers, Agricultural Department and District Forest Officials, Forest Department, Government of Tamil Nadu were also involved in the above interaction. This helped the farmers to discuss their problems at local level and also helped them to appreciate the value of the suggestions made by the researcher and to express their willingness or otherwise to accept the same.

Many of the families in the village both men and women of the household take responsibility for agricultural decisions. Most of the land parcels are family farm and fragmented and sometimes consisting of extremely small parcels, which requires greater efforts for cultivation and the size of land holding (small) thus remains a major hurdle for executing modern practices. The views of the small and marginal farmers are that there has been a considerable hardship in cultivating the land since, most of them depend upon the monsoon. If the monsoon fails, all their farming efforts become a total waste of resources and investment, causing hardships in sustenance. This causes many of the small and marginal farmers migrate in search of other employment, leaving their land uncultivated, to meet their annual subsistence needs. Much of the area is given over to subsistence crops and food security is given prime importance. Crops such as sugarcane, banana, cotton, groundnut and sunflower, which require adequate water supply are widely grown by medium and large farmers than the small and the marginal, who lack capital and are not willing to take risk.

Crop management is largely traditional over the years. There are a few farmers who operate well-managed farms while others have some problems or the other to deal with and often lack managerial capabilities. In principle, many of the farmers including small and marginal farmers tend to agree with the suggestions that require minimum efforts and hesitant when suggesting conservation measures, which is due to lack of fund and proper awareness. In the case of large farmers as discussed above, they are willing to adopt new techniques as well as suggestions to reclaim their hitherto, unproductive culturable land.

Assessing the Level of Acceptability

As stated in the previous section, attribute data of the micro-watershed is used to identify farmers and their land holding capacity, which is used in the assessment of level of acceptability. Farmers are categorized into large, medium, small and marginal based on their land holding capacity and has been interviewed for their acceptability of the suggestions for reclamation activities. Reclamation may be carried out either with minimum effort or by adopting conservation measures (maximum effort) and accordingly these two suggestions have been forwarded to each farmers. Totally, 47 farmers have been selected covering the entire micro-watershed of both the cadastral sites, which consisted of 4 large farmers, 6 medium farmers, 12 small farmers and 25 marginal farmers.

Level of Acceptability in the Micro-watershed

The study area is mostly agriculture predominant. A large proportion of farmers have switched over from cultivating their own lands and have to become agriculture labors in neighboring farms or places. This is mostly observed to be true in the case of small and marginal farmers who for their sustenance depend upon the agricultural related activities. The farmers (large, medium, small and marginal) are interviewed for their acceptability of both the suggestions (with minimum effort and maximum effort). The details of sample size of farmers in each category and their acceptance to suggestion are given in Table3 - 6.

Nearly 68 per cent of the farmers agreed with the first suggestion (reclamation of land with minimum effort) and only 25.5 per cent of them agreed to adopt any conservation measures to reclaim their lands. 23.4 per cent of the farmers agreed to both the first and second suggestion with some reservations. The reservations are based on the factors such as affordability, risks involved, uncertainty of any returns from land after adopting conservation, deviating from traditional practices, societal set up and marketing. 8.5 per cent of the farmers did not agree for the first suggestion and the percentage increase to 51.1 in the case of second suggestion. The refusals to accept the suggestions are mainly due to the economic condition of the local community.

While analyzing the individual categories of farmers, it is observed that 100 per cent of large farmers agreed to the first suggestion. The level of acceptability is relatively lesser in the case of medium farmers (67%), small farmers (75%) and marginal farmers (60%). For the second suggestion, nearly 75 per cent of the large farmers accepted the reclamation activities and to adopt proper conservation measures. Only 50 per cent of medium farmers agreed for the second suggestion and it is still lower in the case of small (33.3%) and marginal (8%) farmers. Medium (33%), small (16.6%) and marginal (28%) farmers showed keen interest regarding the suggestions and accepted the reclamation activities as per the first suggestion with some reservation. Percentage of acceptability remained same in the case of medium farmers for the second suggestion and varied for small (41.6%) and marginal (12%)

Table 6. Level of Acceptability among the Farmers

Farmers withSample Size	Acceptability (Number and Percentage)	
	Suggestion –1	Suggestion – 2
Large Farmers (4 persons)	A – 4 (100%) AR – 0 NA – 0	A – 3 (75%) AR – 01 (25%) NA – 0
Medium Farmers (6 persons)	A – 4 (67%) AR – 2 (33%) NA – 0	A – 3 (50%) AR – 2 (33%) NA – 1 (17%)
Small Farmers (12 persons)	A – 9 (75%) AR – 2 (16.6%) NA – 1 (8.4%)	A – 4 (33.3%) AR – 5 (41.6%) NA – 3 (25.1%)
Marginal Farmers (25 persons)	A – 15 (60%) AR – 7 (28%) NA – 3 (12%)	A – 2 (8%) AR – 3 (12%) NA – 20 (80%)

• A – Accepted, AR- Accepted with sum reservation, NA – Not Accepted

farmers. Many of the marginal farmers are reluctant to accept both the first (12%) and the second (80%) suggestions. This is mainly due to the economic constraints apart from their reluctance to take risks and deviating from traditional agricultural practices. Percentages of not accepting any reclamation activities by small farmers are estimated as 8.4 for the first suggestion and 25.1 per cent for the second suggestion, which again may be attributed to the lack of awareness of the people to conservation techniques and limitation in their affordability. This approach is carried out to understand the awareness of local community and willingness to adopt any conservation activities.

Issues and Policies

The suggestions and recommendations derived from the scientific approach, when carried out at field level, require an understanding of issues and policies pertaining to wasteland developmental activities. Planning must also be adopted considering all these issues and policies for effective utilization of wastelands and a brief description about them are given below.

1. Wastelands suitable for agricultural purposes should be taken up as a major policy issue for increasing the agricultural production.
2. Scientists involved in agricultural and soil conservation activities, using both conventional and remote sensing techniques, should help the farmers, save the land resources available and change the cropping pattern to make better use of their lands in terms of productivity.
3. Voluntary agencies and the Government at local level should assist the poor farmers by conducting demonstrations of improved technical methods, providing appropriate agricultural technology for reclamation, and providing technical and financial support.
4. Local level Government agencies should provide required input such as leasing of land, credit, agricultural technologies, marketing and other required raw materials for agricultural reclamation of wastelands on a subsidized basis to the farmers.
5. Required agricultural technical input should be strengthened and modern technologies for appropriate wastelands reclamation made available to the needy and interested farmers for the reclamation of wastelands.
6. Wider publicity campaigns creating awareness among farmers by means of providing necessary training, education and demonstrating modern techniques to the farmers through various

media like radio, television, local newspapers and films for their involvement.

7.The administrative rules, regulations and procedures should be simplified to help the farmers providing them technical and moral support to reclaim the wastelands.

All the above issues and policies require a proper follow-up action and ensure cultivation in the reclaimed wastelands with a view to improving productivity.

Conclusion

There is a clear distinction in suggestions for reclamation of wasteland categories observed in study area. Since the micro-watershed is agriculture predominant, people are more inclined to growing traditional crops. When conditions such as monsoon, do not favor them, they work as agricultural laborers. At the same time, lack of fund and lack of awareness cause the local community to look for alternatives for subsistence instead of agriculture, leaving their land barren. Though the local farmers agreed to the suggestions, they are reluctant in adopting conservation activities, which involves money and intensive labour. The government and other voluntary organizations should help in restructuring the land use practice and thus avoid further degradation of land.

REFERENCES

Bhavani, V., Shankar and Muthukrishnan, N., (1996). A survey of Wastelands in Tamil Nadu with Reference to Their Nutrient Status. Indian Journal of Environmental Protection, 16 (1), pp. 29 -35.

Douglas, P.S., (1994). Resources Management Perspective: Practical Consideration for using GIS and Remote Sensing at the field level. Remote Sensing and GIS in Ecosystem Management, Washington: Island Press, pp. 140-156.

Dutt, G.K., (1988). Wastelands in India: Their Assessment and Mapping', Wastelands Development and their Utilization. Proceedings of National Seminar held at Central Arid Zone Research Institute (CAZRI), Jodhpur: Scientific Publishers. pp. 1 -17.

Food and Agricultural Organisation of the United Nations, (1992). Report on Agriculture Towards 2000. Rome: Land and Water Development Division, Agriculture Department, pp. 2-3.

Gautam, N.C., Narayan, L.R.A., Chenniah, G.Ch., Ragav Swamy and Nagaraja, R., (1986). Use of Remote Sensing Techniques for Mapping and Monitoring of Wastelands in India, Dimensions of Wastelands Development. Proceedings of National Seminar on Wastelands Development, New Delhi, pp. 89- 98.

Ghosh, P.K. and Mittal, J.P., (1988). Role of Animal Husbandry in Wastelands Management, Proceedings of National Seminar on held at Central Arid Zone Research Institute (CAZRI), Jodhpur: Scientific Publishers, pp. 119-123.

Gibbs, R.J., (1970). Mechanism Controlling the World Water Chemistry, Science, 01.17, pp. 1088-1090.

Hegde, N.G., (1993). Development of Wastelands: Role of Micro-Level Planning, Wasteland News, 9 (1), pp. 40 -45.

Jasbir Singh, (1971). Agricultural Colonization of Cultivable Wasteland in India. The Deccan Geographer, 9 (2), pp. 135-149.

Kulandaiswamy, V.C., (1997). Report of the High Level Committee on Development of Wastelands, Watershed Development and Implementation of Poverty Alleviation Programmes, Chennai: Department of Agriculture, Government of Tamilnadu.

Majid Ghiassi Far, (2001). Land use / Land cover mapping by integrating remote sensing and Geographical Information techniques. ICORG, pp. 99-107.

Montapa, F., (1972). Aerial Photographs and Their Use Lor locating and Mapping Wastelands of India, Journal of the Indian National Science Academy, Bulletin No.44, pp. 59-66.

Mruthyunjaya, S.P., Malhotra, M., and Sen, A.K., (1992). Technical Possibilities of Agricultural Growth and Stability in Western Rajasthan -An Assessment. Annals of Arid Zone, 22 (1), pp. 121-133.

NAEB (1997). Optimal Landuse Planning and Carrying Capacity of the Land Using Remote Sensing and GIS Technique, Unpublished Project Report, New Delhi: National Afforestation and Eco-development Board, Ministry of Environment and Forests, pp. 89.

Nagarathinam, V., (1997). Geographic Information System (GIS) for Land Evaluation and Planning, in Proceedings of the National Seminar on Geographic Information System (GIS) for Development Planning, Chennai: The Ranganathan Centre for Information Studies, pp. 53-63.

Nagarathinam, V. and Hamsanandhan, H., (1993). Pilot Project on Application of Geographic Information System (GIS) for Rural Land Management - Wastelands Development Planning in Usilampatti Block, Madurai District, Tamil Nadu, Project Report New Delhi: National Afforestation Eco-Development Board, Ministry of Environment and Forests, p. 32.

NRSA, (1997). Manual of Procedure for Preparation of Wastelands Digital Database using Remote Sensing and GIS Techniques, Hyderabad: National Remote Sensing Agency. p. 210.

NWDB, (1986). Wasteland: Definition and Database', Report of the Technical Task Group, New Delhi: National Wasteland Development Board, p. 56.

Oberoi, G.S. and Singh, C.B., (1986). Reclamation of Wastelands Information Monitoring: A Case Study of a Part of Karnal District (Haryana), Dimensions of Wastelands Development, Proceedings of National Seminar on Wastelands Development, New Delhi, pp. 99- 110.

Pisharoty, P.R., (1987). Water Management and Wastelands Development, Wasteland News, 1 (1), pp. 3-11.

Rai, P., (1999). Growing Greener and Increase the Production of Wastelands, Wasteland News, 14 (3), pp. 48-50.

Ramamurthy, J., Uday Bhaskar, N., Kameswara Rao, S.V.C. and Perumal, A., (2001). Soilscape Modelling using Geomorphology and Slope Information in a GIS Environment - ICORG 2, pp. 337 -348.

Sankarnarayan, K.A., (1988). Agroforestry: The Key to Rehabilitation of Wastelands, Wastelands Development and their Utilization, Proceedings of National Seminar held at Central Arid Zone Research Institute (CAZRI), Jodhpur, Scientific Publishers, India, pp. 322 -342.

Sen, A.K., (1972). Land Utilization Mapping to Estimate the Wastelands of Arid Zones in Rajasthan by Photo-Interpretation Technique, Indian National Science Academy Bulletin, 44, pp.67-71.

Shah, S.A., (1995). Involvement of Industry in Development of Degraded Forest Lands, Wasteland News, 10 (4), pp. 29- 30.

Shankarnarayan, K.A. and Singh, S., (1979). Application of Landsat Data for Natural Resource Inventory and Monitoring of Desertification Report of the Visiting Int. Scientific Programme, Remote Sensing Institute, SPSU, pp. 184-230.

Sharma, J.P., Rajeev Srivatsava, Tapas Bhattacharya, Shyampura R.L., Saxena, R.K., Gajbhiye, K.S., (2001). Utilization of Remote Sensing Data in Soil Resources Mapping for Land Use Planning - ICORG pp. 270- 274.

Singh, A.L., (1971). Time Distribution and Utilization of Uncultivated land in Kohli Tehsil, Geographical Review of India, 39 (3), pp. 206-211.

Singh, S.N., (1969). Distribution and Changing Pattern of Cultivable Wastes in Kerakat Tehsil, Jaunpur District, Uttar Pradesh, National Geographical Journal of India, 15 (3 and 4), pp. 207-223.

Singh, Y.P. and Mittal, S.P., (1993). Watershed Management Approach to Wastelands Development, Wasteland News, 8 (2), pp. 49- 51.

Sinha, B.P. and Dubey, B.P., (1986). Status, Distribution and Development of Wastelands in Bihar, Dimensions of Wastelands Development, Proceedings of National Seminar on Wastelands Development, New Delhi, pp. 23-30.

Sivasamy, R., Natarajan, S., Ramaswami, C., Mohamed Ali, A. and Palaniswami, (1997). GIS for Farm Level Developmental Planning, Proceedings of symposium on Remote Sensing for Natural Resources with Special Emphasis on Infrastructure Development, Hyderabad: Indian Society of Remote Sensing, pp. 85-89.

Stamp, L.D., (1948). The Land of Britain: Its use and Misuse, London: John Willey and Sons Ltd., p. 433.

Swaminathan, M.S., (1997). Towards a Hunger -Free India, Wastelands News, 13 (1), pp. 8-9.

TAWDEV, (2001). Draft Project Proposal on Wasteland Development, Tamil Nadu Wasteland Development (TAWDEV) Programme, Agriculture Chennai: Department, p. 27.

Vijay, U. and Pradeep, Kumar, A.V., (2001). Micro-Level Wasteland Development on Watershed Basis using GIS and Remote Sensing: A Case Study of Beguruhalla Watershed, Shikaripur, Shimoga District, Karnataka, 2, pp. 152-157.

Wasteland Survey and Reclamation Committee, (1961). Report on the Locational and Utilisation of Wastelands in India, Part-X, Uttar Pradesh, New Delhi: Ministry of Food and Agriculture, pp. 34-97.

Yadav, S.S., (1986). Geographic Aspect of Wastelands Development in India, Dimension of wastelands development, New Delhi: Proceedings of the National seminar on Dimensions of wastelands development, , pp.31-34.

Zinck, A.J. and Valenzuela, C.R., (1990). Soil Geographic Database: Structure and Application Examples, ITC Journal, 3, pp.270-293.

APPLICATIONS OF GEOSPATIAL TECHNOLOGY
A. Ganesh (Ed.), (2006) : pp. 113-131
Satish Serial Publishing House, New Delhi
E-mail: ssp@ndf.vsnl.net.in, hkjain1975@yahoo.com
website: www.satishserial.co

7

Watershed Delineation and Characterization

A. GANESH AND S. ARUCHAMY

A watershed is a well defined hydrological unit. In a variety of ways the watershed has formed a framework for human activity. Indeed, the idea of the watershed as an appropriate aerial unit for the organization of human activity and for regional planning has only been revived with the recognition of a system in which soil and vegetation cover as well as hydrological balance are involved, which needs integrated plans and policies to deal with problems posed by flood control, sedimentation, soil erosion, hydro electric power generation, navigation and even nature conservation and stream pollution [Smith, 1969].

Watershed-wise land and water resource studies are essential pre-requisite for any type of development planning. With a present approach to rural development through micro-level planning by the governments, both at the central and state level, a thorough study on the availability and the pattern of utilization of land and water resources at the watershed level is of utmost important. These studies are essential to help modernize the land and water - use policies. Geographical analysis will assist proper planning of land and water resources. The control and use of land and water for the various uses of man, in watersheds as a whole or in the sub-watersheds are important. This concept has been stated by the United Nations Panel on Integrated River Development [United Nations, 1958]. New dimensions have been added with the realizations of the importance of measures like soil reclamation, flood control and afforestation which go to make river valley development projects truly multipurpose in character [Dayal, 1971].

Remote Sensing and GIS Applications in Watershed Management

Everyone is aware that both land and water are the basic natural resources for biomass production and constitute the core of ecological systems. The watershed approach is more rational because the resources of land and water have optimum interaction and synergistic effect when developed on a watershed basis. Remote sensing has become an inevitable tool for land and water resources inventory at local, regional and global scales. Remote sensing provides solutions to facilitate sustainable development of land and water resources, ecological balance and improve the socio-economic conditions of the concerned watershed communities.

The information derived from satellite data, topographical maps and other socio-economic data could be stored in GIS as a database. GIS enables effective and efficient manipulation of spatial and non-spatial data for scientific management of watershed and develop alternatives of development model for the benefit of local people. It also facilitates modeling to arrive at locale specific solutions by integrating spatial and non-spatial data such as thematic layers and socio-economic data. The spatial database facilitates the authorities in planning and change monitoring and also assists in understanding the effect of developmental activities undertaken by incorporating the data derived from the repetitive coverage of the satellite.

The natural resource base of our nation is under great stress due to the severe biotic pressure and ever intensifying development, as evident from the current scenario and projected demands. Statistics reveal that scientific planning, and management of the natural resources has now become imperative rather an option for sustainable development of our country.

In recent times, the development of land and water resources is being planned on the basis of natural hydrologic entities rather than the administrative boundaries. Such development plans are on a sustained basis without deterioration of the ecology with constant increase in productivity for the mankind. The watershed approach is more rational because the resources of land and water have optimum interaction and synergistic effect when developed on a watershed basis. The remote sensing data has the advantage of providing a synoptic view and large area coverage, which helps in obtaining the proverbial "birds eye view" of the features. Thus, it helps in observing, measuring, mapping and monitoring the Earth's natural resources. Satellite remote sensing, in conjunction with GIS, has proved to be an extremely useful tool for natural resource assessment and management. And, at the same time, it helps in micro level planning and implementation.

Remote Sensing - A Cost Effective Tool

Satellite remote Sensing, due to its inherent advantages, as listed below, has become an inevitable tool for land and water resources inventory at local, regional and global scales.

- Synoptic, repetitive coverage on required scales
- Cost-and time effective
- Accurate database generation
- Multi-spectral nature of data helps greatly in the discrimination and mapping of various features in the watershed
- Helps in the monitoring of various developmental activities being undertaken in the watershed
- Change detection capability

With the availability of high resolution spatial and spectral satellite data, the scope and application of satellite data in combination with the GIS has proved to be extremely useful technology for micro level planning and implementation. Thus remote sensing provides solution to facilitate sustainable development of land and water resources. A holistic and integrated approach is needed for the development of rural areas to ensure ecological balance and improve the socio-economic conditions of the concerned watershed communities. In order to derive maximum benefits out of these resources in a sustainable way, it is preferable to treat the watershed as a single unit. A watershed is made up of its physical and hydrological natural resources as well as human resources.

Watershed planning and management of soil and water conservation programmes require scientific information as an essential input for selecting appropriate conservations measures and making them technically and economically effective. Each watershed has its own problems, potentials, prospects and developmental needs under a set of geophysical, climatic, demography and socio-economic conditions. It is necessary to have information pertaining to potentials and problems of the area and specific needs of the people, to evolve local specific action plans. The development of land and water resources within a watershed is best handled by state-of-art tools. There is a need for correlating the resources inventories with the farmers' socio-economic conditions in a watershed.

Role of GIS

The information derived from satellite data, topographic maps and other socio-economic data could be stored in GIS as a database. GIS enables effective and efficient manipulation of spatial and non-spatial data for scientific management of watershed and develop alternatives of development model for the benefit of local people. It also facilitates modeling to arrive at locale specific solutions by integrating spatial and non-spatial data such as thematic layers and socio-economic data. The database generated will also help the authorities in planning and change monitoring and assists in understanding the effect of developmental activities undertaken by incorporating the data derived from the repetitive coverage of the satellite. A GIS is a composite of computer based decision support tools for the integration of spatial data from different sources and for the analysis, manipulation and display of these data. It is therefore, an excellent tool for the management of large bodies of spatially extensive data with all the advantages of a computer environment: precision, consistency and absence of computational error. This powerful tool holds a very large potential in the field of regional and micro-level spatial planning particularly in micro-watershed planning and management. A GIS can help pull together various types of disparate data such as remote sensing data, census data, records from different administrative bodies, topographical data and field observations to assist researchers, planners, project officers and decision-makers in resource management. Creation of a spatial database is the first step in micro-level planning. This is followed by spatial analysis to help identify problem areas and, finally, the steps towards planning to mitigate problems are taken by marking out action areas. Taking a watershed as the spatial unit of study, appropriate physiographic and morphometric parameters can be taken into account to enable proper micro-watershed management.

Watershed Characteristics

Proper planning of watershed is essential for the conservation for water and land resources and their management for optimum productivity. Characterization and analysis of watersheds and inventories of surface water bodies are prerequisites for this. Remote sensing via aerial and space borne sensors can be gainfully used for watershed characterization and assessing watershed priority evaluation problems, potential and management requirements and periodic monitoring. Remote sensing has brought out substantial improvement in watershed survey and mapping techniques during the last three decades [Baly, 1983].

Water resources of any watershed in a semi-arid region are finite and greatly diversified in space and time. Problem issues associated with water resources development and efficient utilization of varied and too much in nature. On the other hand, pressure from various quarters poses challenges. On the other, agriculture is often on the mercy of weather. Added to these are remorseless floods and droughts as frustrated pursuits, regardly low efficiently of irrigation system, menacing problems of water logging and salinity in command areas, agricultural productivity as a cause for major concern, complex nature command area problems and deteriorating water quality all pour problems in varying dimensions. All these are required to be tackled carefully through systematic approaches involving judicial mix up conventional methods and new efforts like remote sensing for optimal use. The advent of remote sensing technology in recently years has considerably contributed for a better water resources management practices with its unique advantages, particularly of satellite remote sensing of synoptic over view and repetitive process. These can give mere real time information and repetitive basis which is of important to monitoring functions. In contrast, data collection by conventional methods would be time consuming and expensive and very often far out of a real time. Proper blend of remote sensing and conventional methodology can greatly improved water resources management activities.

Many programmes have drawn up of utilization of the Indian Remote sensing satellite data. The areas of data utilization, among others include water resources management, particularly flood management, snow hydrology and watershed characterization studies. Studies on drainage watershed characteristics were contributed by Horton [1945], Strahler [1964 and 1971], Kulkarni et al [1968], Daniel [1981], Chang [1982] and Krishnayya and Angira [2002] Gregory and Walling [1976] had studied the geomorphologic approach of the watersheds.

Watershed: Definition and Delineation

The concept of a watershed is basic to all hydrologic designs. Since large watersheds are made up of many smaller watersheds, it is necessary to define the watershed in terms of a point. This point is usually the location at which the design is being made and is referred to as the watershed "outlet". With respect to the outlet, the watershed consists of all land area that "sheds" water to the outlet during a rainstorm. Using the concept that "water runs downhill", a watershed is defined by all points enclosed within an area from which rain falling at these points will contribute water to the outlet.

Delineation Process

The fundamental source of data that we use for delineating and studying watersheds is the Survey of India topographical maps. These maps give a wealth of information including topographic contour lines, locations of settlements, buildings, roads, road types, railroads, pipelines, water bodies, forested land, stream networks, and stream gauging stations and benchmarks.

These maps typically have a scale of 1:50,000. The contour intervals of the elevation data are 20 meter. For watershed delineation, 1: 50,000 sheet maps offer us the best starting point. More detailed analysis would require a detailed topographic survey of the area of interest. Topographic maps are generally available at all Survey of India offices.

Digital Elevation Models

In this age of computers, geographic data can now be stored electronically. Digital Elevation Models (DEM) store topographic data in the form of grid cells. Typically, these grid cells have a resolution of 20 meters and elevation intervals of 1 meter. Using a DEM within a Geographical Information System (GIS), we can perform digital terrain analysis (DTA) such as calculating slopes, flow lengths, and delineate watershed boundaries and stream networks [Figure 1].

Delineation Steps

There are two basic steps to follow in watershed delineation.

Step1 : Choose the point of the watershed outlet. This is generally our point of interest for designing a structure or monitoring location.

Step2 : Delineate the watershed boundary by drawing perpendicular lines across the elevation contour lines for land that drains to the point of interest.

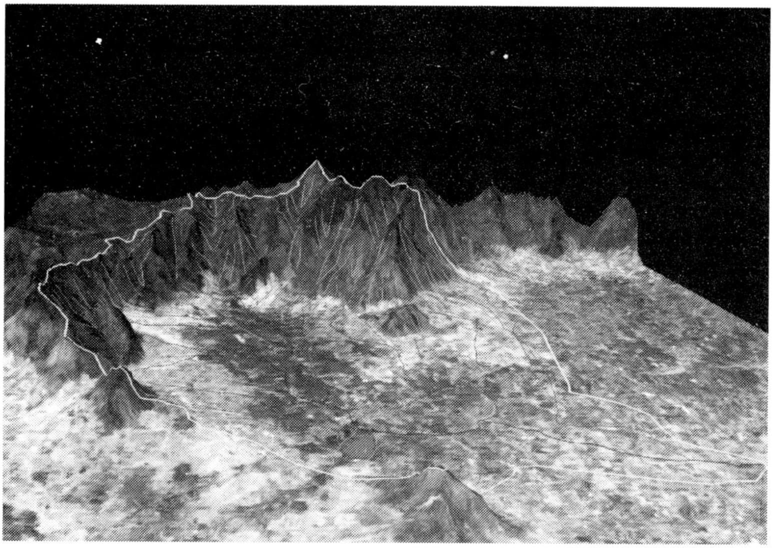

Fig.1 Digital Elevation Model of Kombai Watershed

Note - There are a few things to remember when you are working with topographic maps.

a. A watershed boundary always runs perpendicular to the contour lines.

b. "Arrows" that point upstream are valleys.

c. "Arrows" that point downstream are hills.

Watershed Characteristics

Linear Aspects

Stream Order

The first step in drainage watershed analysis is designation of stream orders. The information of stream order number is useful in relating to the size of its contributing watersheds. Out of the four systems of ordering streams that are available, Strahler's system, which was in fact of slightly modified version of Horton's, was followed because of its simplicity. The smallest unbranched stream segment was designated as the first order stream, the one formed by merging of two such first order segments as the second order stcdfryream, the one formed by the merging of two such second order streams segments as the third order stream and so forth. The trunk stream through which all the dischargc of water and sediment passes is therefore, the stream segment of the highest order. For a watershed, the principal order is defined as the order of the principal channel. The Figure 2 below gives an example of stream ordering.

The concept of stream order is used to compute other indicators of drainage character.

Stream Numbers

After the drainage network elements have been assigned their order numbers, the segments of each order will be counted and the number N_u of the stream segments present in that order is found out. The number of streams decreases as the stream order increases.

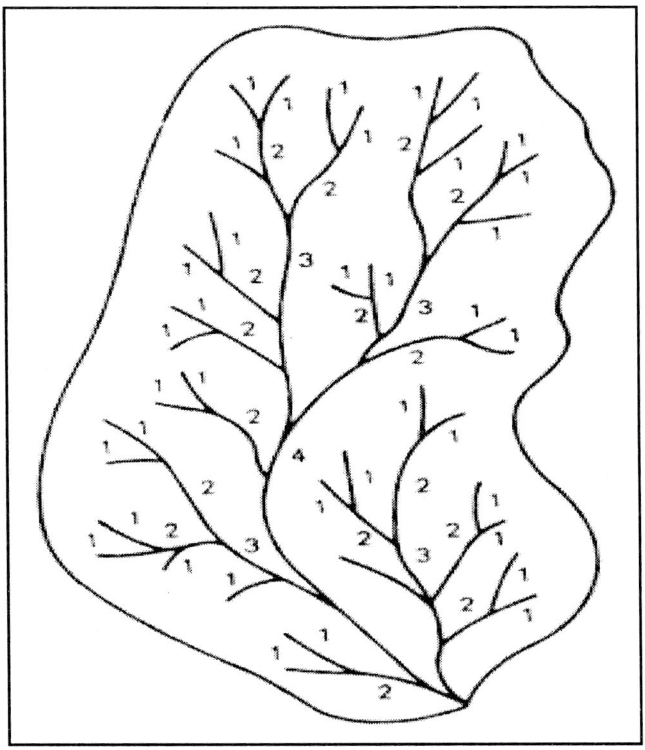

Fig.2 Stream Ordering

Bifurcation ratio

The bifurcation ratio R= N_u/ N_u+1. The bifurcation ratio is calculated by relating the number of segments of a given order N_u to the number of segments of the next higher order N_u+1. From these the mean ratios will be computed. With the actual number of streams that are involved in the above computed ratios, weighted mean bifurcation ratios are also computed. The total number of streams that are involved in the computed bifurcation ratios will be multiplied by the total of the actual number of streams involved in the ratios [Figure 3].

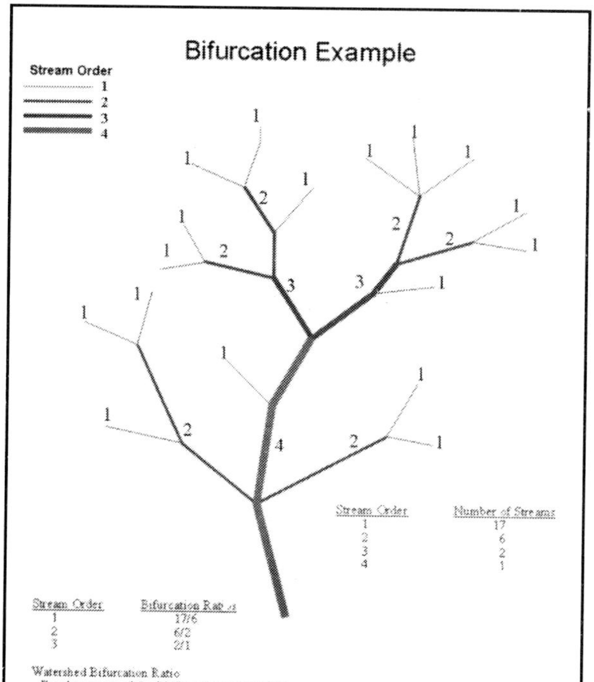

Fig.3 Bifurcation Ratio

Stream Length

The measurement of stream length is essential for computing unit hydrograph log. The length of the various stream segments are measured using a curvimeter. Because all segments of a given order within the specified drainage network are measured successively without pause for recording. Cumulative length appeared on the dial of the curvimeter. The mean length I_u for a given order is obtained by dividing the total length by the total number of segment 'N_u' of that order 'V'. The length ratio, which is the ratio of the mean length of the streams of a given order to the mean length of the streams of a given order to the mean length of the streams of the next lower order, was then computed for each pair of orders [Figure 4 and Table 1].

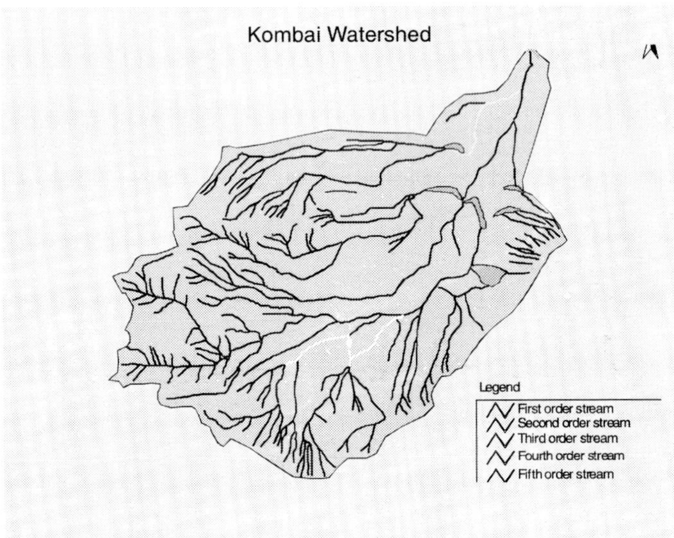

Fig.4 Stream Ordering of Kombai Watershed

Aerial Apects

Drainage Area

The drainage area (A) is the probably the single most important watershed characteristic for hydrologic design. It reflects the volume of water that can be generated from rainfall. It is common in hydrologic design to assume a constant depth of rainfall occurring uniformly over the watershed. Under this assumption, the volume of water available for runoff would be the product of rainfall depth and the drainage area. Thus the drainage area is required as input to models ranging from simple linear prediction equations to complex computer models.

Watershed Length

The length (L) of a watershed is the second watershed characteristic of interest. While the length increases as the drainage increases, the length of a watershed is important in hydrologic computations. Watershed length is usually defined as the distance measured along the main channel from the watershed outlet to the watershed divide. Since the channel does not extend to the watershed divide, it is necessary to extend a line from the end of the channel to the watershed divide following a path where the greatest volume of water would travel. The straight-line distance from the outlet point on the watershed divide is not usually used to compute L because the travel distance of floodwaters is conceptually the length of interest. Thus, the length is measured along the principal flow path. Since it will be used for hydrologic calculations, this length is more appropriately labeled the hydrologic length.

While the drainage area and length are both measures of watershed size, they may reflect different aspects of size. The drainage area is used to indicate the potential for rainfall to provide a volume of water. The length is usually used in computing a time parameter, which is a measure of the travel time of water through a watershed.

Watershed Slope

Flood magnitudes reflect the momentum of the runoff. Slope is an important factor in the momentum. Both watershed and channel slope may be of interest. Watershed slope reflects the rate of change of elevation with respect to distance along the principal flow path. Typically, the principal flow path is delineated, and the watershed slope (S) is computed as the difference in elevation (DE) between the end points of the principal flow path divided by the hydrologic length of the flow path (L):

$$S = DE/L$$

The elevation difference DE may not necessarily be the maximum elevation difference within the watershed since the point of highest elevation may occur along a side boundary of the watershed rather than at the end of the principal flow path.

Watershed Shape

Watershed shape is not usually used directly in hydrologic design methods; however, parameters that reflect watershed shape are used occasionally and have a conceptual basis. Watersheds have an infinite variety of shapes, and the shape supposedly reflects the way that runoff will "bunch up" at the outlet. A circular watershed would result in runoff from various parts of the watershed reaching the outlet at the same time. An elliptical watershed having the outlet at one end of the major axis and having the same area as the circular watershed would cause the runoff to be spread out over time, thus producing a smaller flood peak than that of the circular watershed.

A number of watershed parameters have been developed to reflect watershed shape. The following are a few typical parameters:

1. Length to the center of area (L_{ca}): the distance measured along the main channel from the watershed outlet to the point on the main channel opposite the center of area.
2. Shape Factor (L_l)
   $L_l = (LL_{ca})^{0.3}$
   Where L is the length of the watershed.
3. Circularity ratio (F_c):
   $F_c = P/(4pA)^{0.5}$
   Where P and A are the perimeter and area (ft^2) of the watershed, respectively.
4. Circularity ration (R_c):
   $R_c = A/A_o$
   Where A_0 is the area of a circle having a perimeter equal to the perimeter of the watershed.
5. Elongation Ration (R_e):
   $R_e = 2/L_m(A/p)^{0.5}$
   Where L_m is the maximum length (ft) of the watershed parallel to the principal drainage lines.

Generally, the shape factor (L_l) is the best descriptor of peak discharge. It is negatively correlated with peak discharge (i.e. as the L_l decreases, peak discharge increases).

Channel Length

In addition to the drainage area and the watershed length, the channel length is used frequently in hydrologic computation. Two computational schemes are used to computer the channel length:

1. The distance measured along the main channel from the watershed outlet to the end of the channel, which is denoted as L_c.
2. The distance measured along the main channel between two points located 10 and 85 per cent of the distance along the channel from the outlet, which is denoted at L_{10-85}.

These definitions along with the watershed length are illustrated below. The watershed length is defined by extending a line on the map from the end of the main channel to the divide. This requires some subjective assessment and is often a source of inaccuracy. The definitions for channel length also involve a measure of subjectivity because the endpoint of the channel is dependent on the way the map was drawn.

Channel Slope

The channel slope can be described with any one of a number of computation schemes. The most common is

$$S_c = DE_c / L_c$$

In which DE_c is the difference in elevation between the points defining the upper and lower ends of the channel and L_c is the length of the channel between the same to points. The 10-85 slope can also be used:

$$S_{10\text{-}85} = DE_{10\text{-}85} / L_{10\text{-}85}$$

For cases where the channel slope is not uniform, a weighted slope may provide an index that better reflects the effect of slope on the hydrologic response of the watershed.

Drainage Density

The drainage density is the ratio of the total length of streams within a watershed to the total area of the watershed; thus D has units of the reciprocal of length (1/L). A high value of the drainage density would indicate a relatively high density of streams and thus a rapid storm response. Values typically range from 1.5 to 6 mi/mi^2.

$D = L_t / A$

Stream Frequency

It is defined as the number of stream segments per unit area that decreases with stream order. This is a measure of drainage network and hence the dissection of watershed. More the value of stream frequency, less will be the permeability of the watershed $d_{d \text{ and }} f_d$ are useful in classifying the drainage watershed into various permeability classes which could help decide areas suitable for sinking of wells. Impermeable surfaces can be identified and assessed in infrared images on account of their low thermal inertia (Barrent and Curtis, 1982).

Horton (1945) introduced stream frequency F as the number of stream segments per unit area. It is obtained by dividing the total number of streams (N_{u}) by the total drainage watershed area (A_{u}).

Length of Overland Flow

Horton (1945) used this term to refer to the length of the run of the rain water on the ground surface before it gets localized into definite channels. Since this length of overland flow, on an average is about half the distance between the stream channels, Horton for the sake of convenience had taken it to be roughly equal to half the reciprocal of the drainage density.

Constant of Channel Maintenance

The constant of channel maintenance is the reciprocal of d_d i.e. the area of the watershed divided by the sum of the length of the permanent stream. It shows the area required to maintain each unit length of the stream. Shumm used the inverse of drainage density as a property termed constant of channel maintenance C.

Relief Aspects

Channel Gradient

The stream elevation can be estimated from the contour crossings on the topographic sheets. The total drop in elevation from the source to the mouth can be found out for the elevation from the source to the mouth for the tributaries and the horizontal distances can be measured along the channel using a map measurer.

Study Area: Kombai Watershed

In order to study the watershed delineation and characteristics, the Kombai watershed of the Kambam Valley, Theni district, Tamil Nadu has been selected. The Survey of India topographic sheets at a scale of 1:50,000 and IRS 1C data are used for analyzing watershed delineation and characterization of the Kombai watershed. With the help of the Survey of India topographic sheets, all the drainages including first order streams are traced and watershed of Kombai has been delineated. The drainages from the IRS 1C of the same area has been digitized and overlayed with the help of ArcGIS 9.1. One of the most important applications of Digital Elevation models is watershed delineation. ArcGIS has built-in functions for watershed delineation using DEMs. However, the ArcGIS and its Spatial Analyst extension provide tools which are more convenient to use. These tools have been further developed by the authors into a watershed characterization system of GIS data for hydrologic models.

Figure 5 shows the drainages of the study area using Survey of India topographic sheets and Figure 6 depicts the drainages and micro-watersheds of the same area in IRS 1C images. After the delineation process, different characteristics of morphometry of the Kombai watershed have been calculated based on the procedure discussed in this paper.

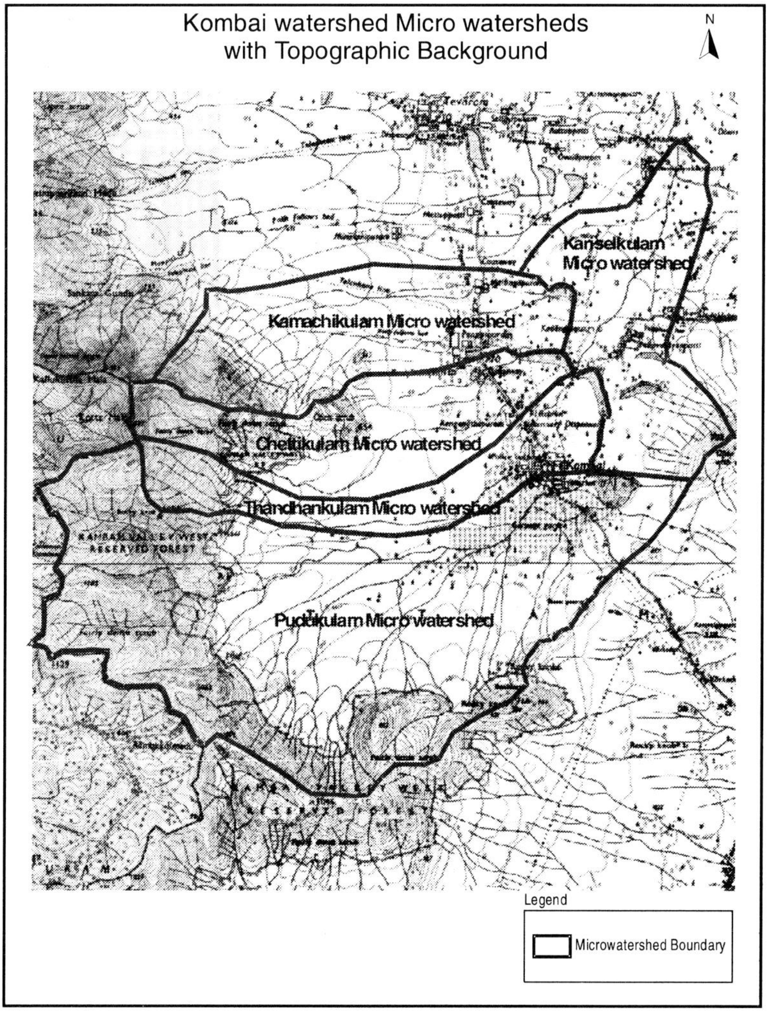

Fig.5 Micro-watersheds of Kombai watershed with Topographic Background

Conclusions

The measurement of these watershed characteristics is important because they are intimately related to watershed water yield and sediment load. The physiographic parameters measurable from space platform include stream order, stream length, watershed area, watershed shape, drainage density, drainage pattern and stream sinuosity. Further water bodies and irrigated and unirrigated areas can also be delineated fr.om high resolution satellite imagery and are valuable in characterizing the hydrologic response of a watershed.

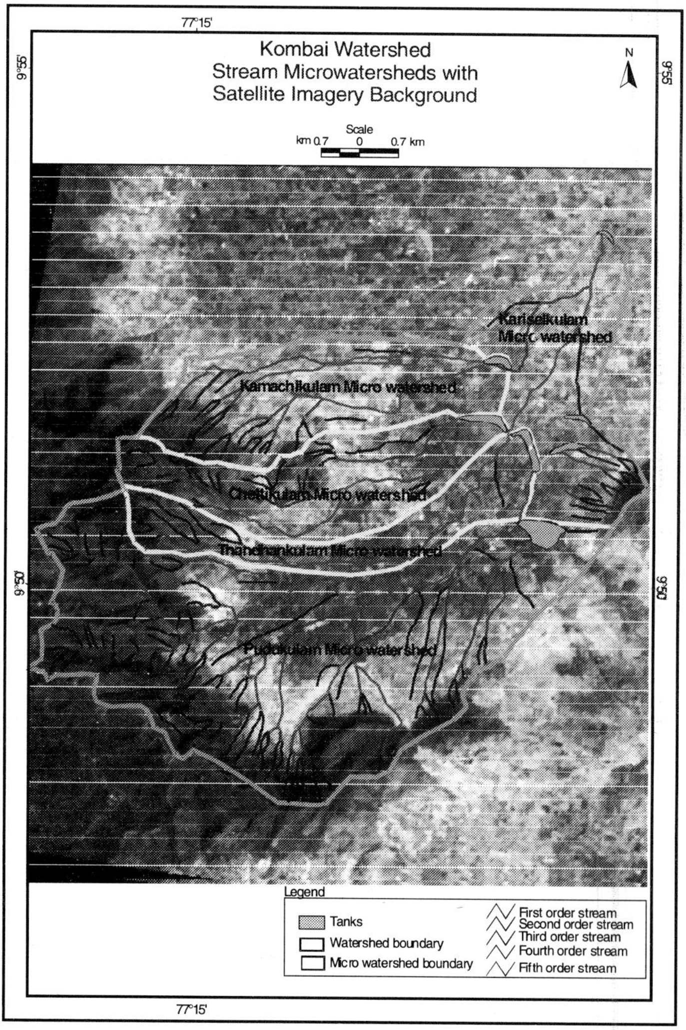

Fig.6 Streams in micro-watersheds of Kombai Watershed with Satellite Image Background

Drainage network information from topographic maps is generally dated. Remote sensing due to its more current nature can often provide better and more current nature can often provide better and more reliable information.

Table 1. Morphometric Characteristics of Kombai Watershed

Name of Micro watershed		Pudukulam Micro watershed	Thandhankulam Micro watershed	Chettikulam Micro watershed	Kamachikulam Micro watershed	Kariselkulam Micro watershed	Watershed as a whole
Stream order		V	V	III	IV	V	V
Micro watershed area in Sq.Km		30.44	6.23	8.69	9.98	8.76	64.10
Stream order	I	87	8	23	19	15	151
	II	21	1	5	7	4	40
	III	8	3	1	3		12
	IV	3			1	1	4
	V	1				1	1
	Total	150.44	12	29	30	21	208
Stream Length in Km	I	53.94	6.92	12.33	11.71	10.37	95.27
	II	17.80	2.19	6.65	6.05	2.04	34.72
	III	14.77	6.75	6.45	10.45		38.47
	IV	4.45			0.43	2.47	7.35
	V	3.64	1.86			4.12	9.63
	Total	94.6	17.72	25.43	28.64	19	185.44
Micro watershed perimeter in Km		28.92	18.11	16.16	15.47	15.85	36.82
Basin length in Km		10.60	7.36	6.94	6.71	5.25	13.10
Mean Stream Length in Km	I	0.62	0.87	0.54	0.62	0.69	0.63
	II	0.85	2.19	1.33	0.86	0.51	0.87
	III	1.85	2.25	6.45	3.48		3.21
	IV	1.48			0.43	2.47	1.84
	V	3.64				4.12	9.63
Stream length Ratio	II / I	0.33	0.32	0.54	0.52	0.20	0.36
	III / II	0.83	3.08	0.97	1.73	0.00	1.11
	IV / III	0.30			0.04		0.19
	V / IV	0.82				1.67	1.31
Elongation Ratio		0.59	0.38	0.48	0.53	0.64	0.69
Drainage Texture Ratio		5.20	0.66	1.79	1.94	1.33	5.65
Bifurcation Ratio	I / II	4.14	8.00	4.60	2.71	3.75	3.78
	II / III	2.63	0.33	5.00	2.33		3.33
	III / IV	2.67			3.00	0.00	3.00
	IV / V	3.00				1.00	4.00
Mean Bifurcation Ratio		3.11	4.17	4.80	2.68	1.58	3.53
Drainage Density Km/Sq.KM		3.11	2.85	2.93	2.87	2.17	2.89
Stream Frequency		4.94	1.93	3.34	3.01	2.40	3.24
Form Factor		0.27	0.11	0.18	0.22	0.32	0.37
Circularity Ratio		0.46	0.24	0.42	0.52	0.44	0.59
Length of overland Flow		0.16	0.18	0.17	0.17	0.23	0.17

The information on drainage network is largely a reflection of the litho logy and structure of the watershed. Stream orders, stream length, stream frequency, bifurcation ratio, stream sinuosity, drainage density and linear aspects of channel system all could be obtained especially using topographic sheets and satellite images.

The quantitative analysis of drainage networks enables relationships between different aspects of the drainage pattern of the watershed to be formulated as general laws and to define certain useful properties of drainage watershed in numerical forms. The laws of stream number, stream lengths and stream slops can be derived from measurements made in the drainage watershed.

REFERENCES

Balakrishnan, P., (1986). Issues in water resources development and management and the role of remote sensing. Bangalore: Indian Space Research Organization.

Bali, Y.P., (1983). Proceedings of the National Symposium on Remote Sensing Development and Management of Water Resources, Indian Society of Photo Interpretation and Remote Sensing. Dehra Dun. pp. 10-14.

Chang, J., (1982). Stream Order and Stream Net Ratios of Drainage Watersheds in Taiwan, Geographical Research, 8, pp.113.156 (In Chinese and abstract in English).

Daniel, J.K.K., (1981). Drainage Density as an Index of Climatic Geomorphology, Journal of Hydrology, 50, pp.147-154.

Gregory, K.J. and D.E. Walling,(1976). Drainage Watershed Form and Process: A Geomorphologic Approach. London: Edward Arnold.

Horton, R.E., (1945). Erosional Development of Streams and their Drainage Watersheds: Hydro Physical Approach to Quantitative Morphology, Geological Society of American Bulletin, 56, pp.275-3730.

Kesik, A.B., (1973) Analysis of the drainage pattern of selected areas of Canada using ERTS-1 Imagery as a base in Remote Sensing and Water Resource Management. American Water Resource Association, pp.209-24.

Krishnayya, J. G., Baruah Angira, (2002). Use of Geographic Information System (GIS) tool in watershed hydrology and irrigation water management. http://www.gisdevelopment.net/application/nrm/water/watershed/watws0004.htm

Kulkarni, V.M., Dehai and S.M, Dorges, (1968). Amba Rover: Drainage Watershed Characteristics, Bombay Geographical Magazine, 14 (1), pp.35-38.

Smith, C.T., (1969). The Drainage Watershed as an Historical basis for Human Activity, in Chorley, R.J., (1969) Water, Earth and Man. London: Methuen and Co. Ltd.

Strahler, A.N., (1964). Quantitative Geomorphology of Drainage Watersheds and Channel Networks, in Chow V.T., (Ed.), (1964). Hand Book of Applied Hydrology, New York: McGraw-Hill Book Company.

Shimano, Y., (1978). On the Characteristics of the Channel Net Works in Japanese Drainage Watershed, Geographical Review of Japan, 51 (10), pp. 776.784.

Strahler, A.N., (1971) Physical Geography, New Delhi: Willey Eastern Private Limited.

United Nations,(1958). Integrated River Watershed Development, United Nations: Department of Economic and Social Affairs.

APPLICATIONS OF GEOSPATIAL TECHNOLOGY
A. Ganesh (Ed.), (2006) : pp. 133-150
Satish Serial Publishing House, New Delhi
E-mail: ssp@ndf.vsnl.net.in, hkjain1975@yahoo.com
website: www.satishserial.co

8

Application of GIS in Hydrological Analysis

K. RUTHARVEL MURTHY AND A. GANESH

Hydrology is the science that encompasses the occurrence, distribution, movement and properties of the waters of the Earth and their relationship with the environment within each phase of the hydrologic cycle. The hydrologic cycle is a continuous process by which water is purified by evaporation and transported from the Earth's surface including the oceans to the atmosphere and back to the land and oceans. All of the physical, chemical and biological processes involving water as it travels it various path in the atmosphere, over and beneath the Earth's surface and through growing plants, are of interest to those who study the hydrologic cycle. There are many pathways the water may take in its continuous cycle of falling as rainfall or snowfall and returning to the atmosphere. It may be captured for millions of years in polar ice caps. It may flow to rivers and finally to the sea. The cycle for water may be short, or it may take millions of years. Water is diverted temporarily from one part of the cycle by pumping it from the ground or drawing it from a river or lake. It is used for a variety of activities such as households, businesses and industries; for transporting wastes through sewers; for irrigation of farms and parklands; and for production of electric power.

GIS is focused on representing the landscape by means of locationally referenced data describing the character and shape of geographic features. A spatial hydrology model is one, which simulates the water flow and transport on a specified region of the Earth using GIS data structures. A fundamental distinction arises in the treatment of spatial units depending upon whether raster or vector data structures are used. In a raster data structure, a fine mesh of cells is laid over the landscape and all calculations are done for each cell. This approach, based on using digital elevation model cells as the spatial units, is very useful for certain kinds of hydrologic analysis. The vector data structures are based on related to points, lines and polygons, mostly this type of data structures are constructed using X, Y Cartesian coordinate system.

Hydrological Analysis

The last few years have witnessed an enormous interest in application of GIS in hydrology. The GIS technology has the ability to capture, store, manipulate, analyze and visualize the geo-referenced data. On the other hand hydrology is inherently spatial and distributed hydrologic models have large data requirements. The integration of hydrology and GIS is there for quite natural. The integration involves three major components: 1. Spatial data construction, 2. Integration of spatial model layers, 3. GIS and model interface. Hydrological modeling is a powerful technique in the planning and development of integrated approach for management of water resources.

GIS can assist in design, calibration, modification and comparison of models. GIS has influenced the development and implementation of hydrologic models at several different levels, and it provides representations of these spatial features of the Earth, while hydrologic modeling is concerned with the flow of water and its constituents over the land surface and in the subsurface environment. GIS and hydrological modeling can be considered as complimentary. GIS could benefit from the temporal modeling capabilities of hydrological models and hydrologic models can benefit from the spatial modeling capabilities of GIS. GIS with its upcoming advanced technology has been a great asset to hydrologic modeling.

Review of Literature

Seth and Dee (1993) have conducted studies on hydrology by using a GIS – based solution for watershed analysis for Maryland state. They constructed a model named as GISHYDRO. The use of GISHYDRO saved the project a significant amount of time by automating the process of developing RCN numbers, drainage areas, slopes, stream lengths and times of concentration. Miller and Semmens (1995) have developed automated geo-spatial watershed assessment tool (AGWA) in GIS for analyzing the water resources. The key components of AGWA are the hydrological models used to evaluate the effects of land cover and landuse on water response. Poiani, et.al.,(1995) were used GIS to model non-point source (NPS) pollution in watersheds with wetlands using both coupled model and modeling within a GIS. The National Institute of Hydrology, has been carrying out research on various aspects of hydrology and water resources using remote sensing and GIS techniques. These studies cover different hydrologic processes occurring in a catchment e.g. rainfall-runoff modeling, soil erosion, reservoir sedimentation, land use/cover and land capability studies, watershed studies, snow and glacier melt etc.

In a study on watershed modeling with GIS based distributed unit hydrograph approach, a spatially distributed unit hydrograph for Temur watershed at railway bridge no. 293 (M. P.) has been developed. The method of distributed unit hydrograph computation allows for spatial non-uniformity of excess rainfall. Consequently, it is based on the time-area method derived using GIS. The GIS allows development of a watershed's channel network for calculation of realistic travel times, it handles the distributed excess rainfall in calculating local surface runoff rates as inputs for channel flow and it compiles the time-area diagram from which distributed unit hydrograph is derived.

Aim

The main aim of the study is to investigate the hydrological conditions of the Vellar basin Tamil Nadu using GIS. The objectives are to explore the applications and aspects of hydrological tool in ArcGIS and to derive the hydrological information from the investigations.

Selection of the Study Area

Vellar basin has been selected for the study of hydrological analysis. This basin is characterized by diverse topography. Rugged topography has been noticed over western, southwestern and northwestern areas and the eastern portions are covered by plain topography. The western region is dominated by many small originating streams, which attributed to the diverse character of streams draining on this region. These streams join together and they form a major river system, which flows towards east from western portion of this basin. So, it provides a single frame of watershed, which can be analyzed using modern technology (GIS) to make hydro planning for basin.

Hydrological Applications

The shape of the Earth's surface is useful for many fields such as regional planning, agriculture, and forestry. These fields require an understanding of how water flows across an area, and how changes in that area may affect that flow. When modeling the flow of water, one may want to know where the water came from and where it is going. For this hydrologic analysis functions helps to create models and to know the movement of water across a surface. The area upon which water falls, and the network through which it travels to an outlet is referred to as a drainage system. The flow of water through a drainage system is only a subset of what is commonly referred to as the hydrologic cycle, which also includes precipitation, evapotranspiration, and groundwater.

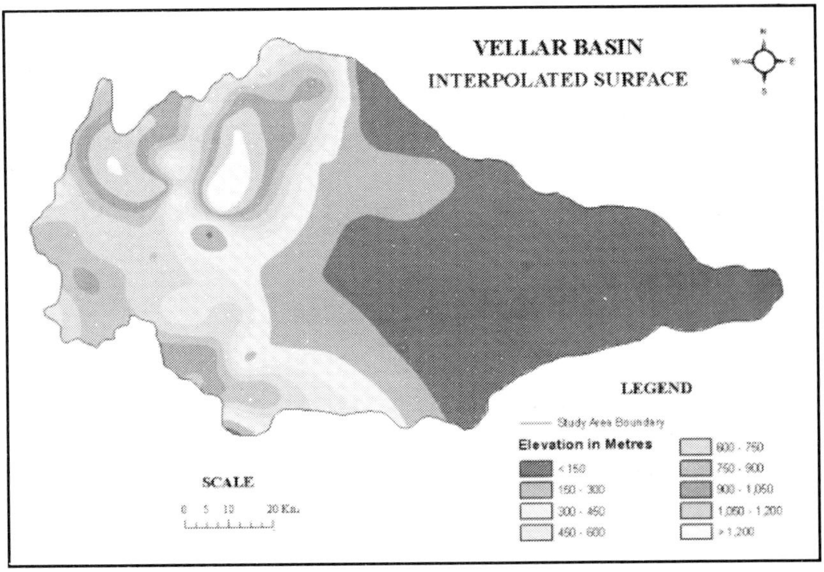

Fig 1. Interpolated Surface

A drainage basin is an area that drains water and other substances to a common outlet. Other common terms for a drainage basin are watershed, catchment, or contributing area. This area is normally defined as the total area flowing to a given outlet, or pour point. An outlet or pour point is the point at which water flows out of an area. This is usually the lowest point along the boundary of the drainage basin. The boundary between two basins is referred to as a drainage divide or watershed boundary. A watershed is the upslope area contributing flow to a given location. Such an area may also referred to as a basin, catchment, sub watershed, or contributing area. A sub watershed is simply part of a hierarchy implying that a given watershed is part of a larger watershed. Watersheds can be delineated from a DEM by computing the flow direction and using it in the Watershed function. The Watershed function uses a raster of flow direction to determine contributing area. Interpolation predicts values for cells in a raster from a limited number of sample data points. It can be used to predict unknown values for any geographic point data like elevation, rainfall and so on.

Interpolated surface has been created as per the given Spot heights, Bench marks and Triangulation stations by using spline interpolation method and that estimates values using a mathematical function that minimizes overall surface curvature, resulting in a smooth surface that passes exactly through the input points. The Regularized method creates a smooth, gradually changing surface with values that may lie outside the sample data range. By using the interpolated surface slope, aspect and TIN surfaces could be created by using spatial analyst tool.

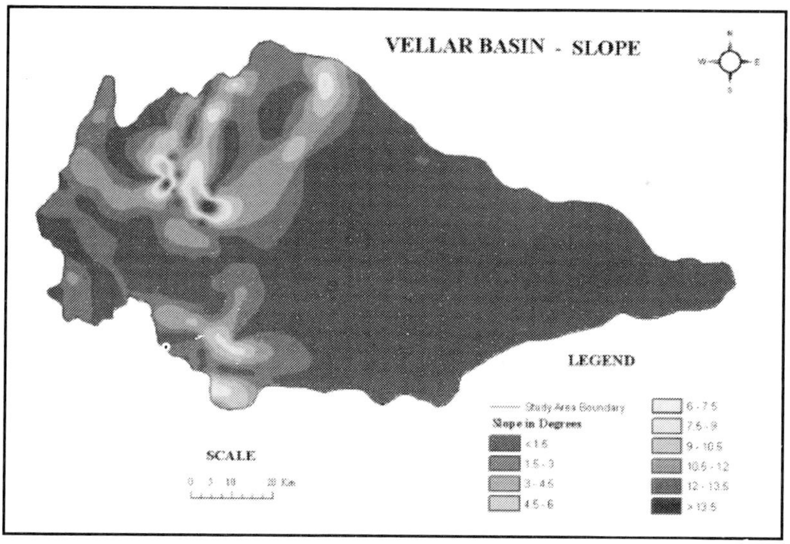

Fig 2. Slope

Slope is the steepness or gradient of a unit of terrain, usually measured as an angle in degrees or percentage. Aspect refers to the direction a slope faces. That is, if one were standing on a hillside and looking down the slope, the direction you were facing would be the aspect of the slope. It is usually broken into eight categories: north, northeast, east, southeast, south, southwest, west, and northwest.

The Triangulated Irregular Network (TIN) is a data model that attempts to incorporate the topography of mapped features. It consists of a multitude of nodes with known elevations, which are connected by line segments (arcs) to other nodes to form triangles. These triangles represent areas of uniform slope and can be various sizes. The database keeps track of the *topology* of the triangles, or how they are connected to one another. A TIN model can be used to calculate *flow directions* and watershed areas, as well as a variety of other applications.

Hydrological Tools in ArcGIS

ArcGIS Spatial Analyst contains specialized tools for working with and deriving new information from hydrologic and landscape information. When modeling the flow of water, one can know where the water came from and where it is going. ArcGIS Spatial Analyst provides tools to calculate flow across an elevation surface, which provides the basis for creating stream networks and watersheds, calculating flow path length, assigning stream orders, and so forth. These kinds of derived data are often used to aggregate landscape information for input to hydrologic models.

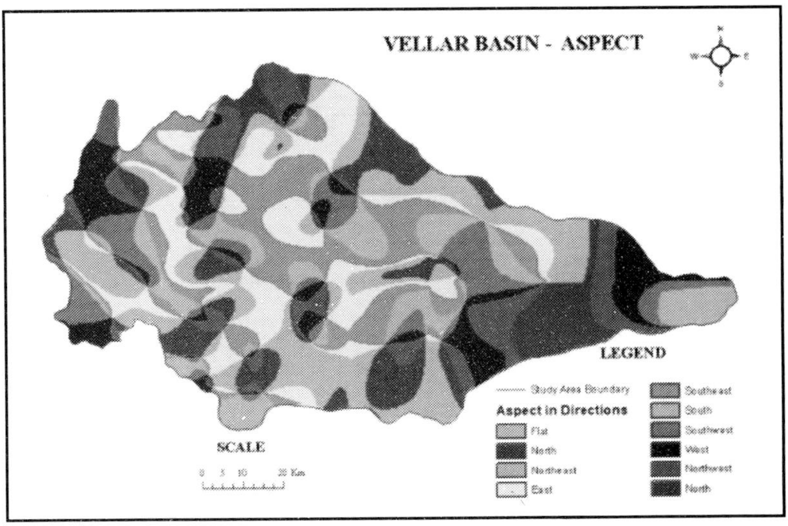

Fig 3. Aspect

Basin

The Basin function delineates drainage basins within the analysis window by identifying ridgelines between basins. Basin analyzes the flow direction raster to find all sets of connected cells that belong to the same drainage basin. The drainage basins are created by locating the pour points at the edges of the analysis window where water would pour out of the raster, as well as sinks, then identifying the contributing area above each pour point. This results in a raster of drainage basins. The basin function gives the result by delineating the major basin into several sub-basins. Here the Vellar basin has been resulted with 78 sub-basins and these sub-basins has been reclassified and generalized for getting convenient results. The sub-basin labeled 8 is the major sub-basin among the sub-basins present in the study area. The entire vellar river comes under this basin and situated in the middle part of the study area. The topography is the main phenomenon, which decides the basin extent. The western portion of the study area has many tiny sub-basins; this may be due to the rugged topography of this region and also having many number of first and second order streams. The middle and eastern portions come under plain region. This region has only few sub basins comparing to western part of the study area.

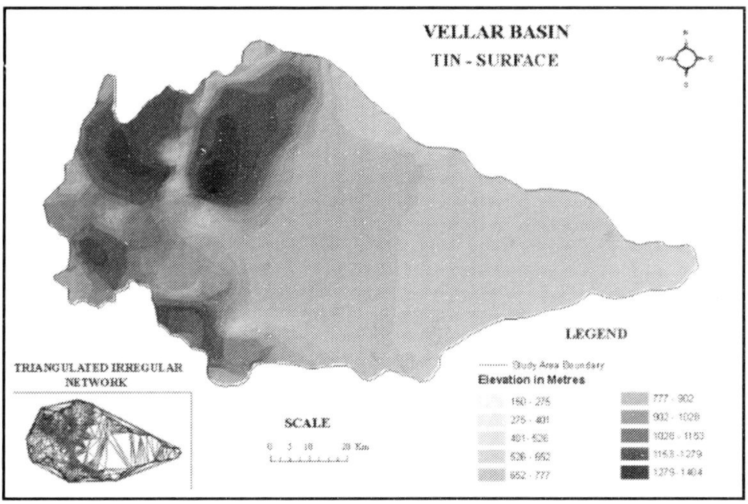

Fig 4. TIN Surface

Sink

A sink is a cell or set of spatially connected cells whose flow direction cannot be assigned one of the eight valid values in a flow direction grid. This can occur when all neighboring cells are higher than the processing cell, or when two cells flow into each other creating a two-cell loop. Sinks are considered to have undefined flow directions and are assigned a value that is the sum of their possible directions. Sinks are considered to have undefined flow directions and are assigned a value that is the sum of their possible directions. For example, if the steepest drop and, therefore, flow direction, is the same both to the right (1) and left (16), the value 17 would be assigned as the flow direction for that cell.

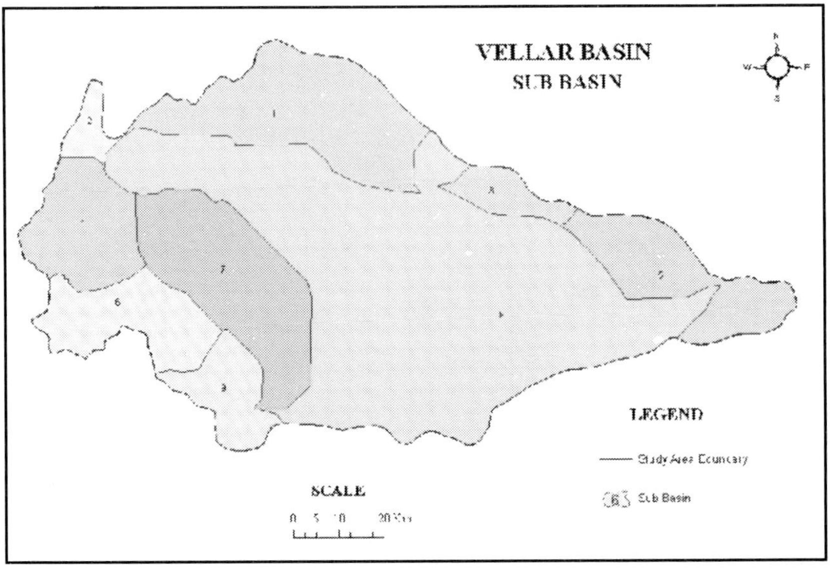

Fig 5. Sub basin

To create an accurate representation of flow direction and therefore, accumulated flow, it is best to use a dataset that is free of sinks. A digital elevation model that has been processed to remove all sinks is referred to as a depression less DEM.

The out put of sink option results by having blacks spots over the elevated surface. These black spots and dots are pixels values bearing noise values, which create a discontinuous surface along the river course. These sinks are emerged due to the error encountered while creating an elevation model and this can be over come by fill option. The sink map shows maximum number of sinks over the hilly terrain because of difficulty encounters while generating a DEM.

Fill

The Fill function fills the sinks in a surface raster to remove small imperfections in the data. Sinks and peaks are often errors due to the resolution of the data or rounding of elevations to the nearest integer value. Sinks should be filled to ensure proper delineation of basins and streams. If the sinks are not filled, a derived drainage network may be discontinuous.

A sink is a cell with undefined drainage direction; no cells surrounding it are lower. The pour point is the boundary cell with the lowest elevation for the contributing area of a sink. If the sink were filled with water, this is the point where water would pour out. Fill uses the equivalents of several functions such as Focal Flow, Flow Direction, Sink, Watershed and Zonal Fill to locate and fill sinks. The resultant fill map shows that the sinks are disappeared and the discontinuous flow is rectified. The continuous surface has been seen and this will be useful for delineating the streams and its basins. While comparing fill map with sink map the errors laid in the western portions are restored.

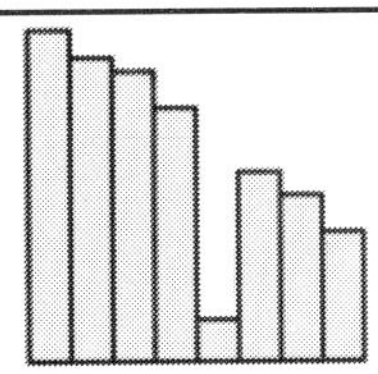

Profile view of a sink

Fig 6.

Flow Accumulation

The Flow Accumulation function creates a raster of accumulated flow to each cell. The results of Flow Accumulation can be used to create a stream network by applying a threshold value to select cells with a high-accumulated flow. Cells of undefined flow direction will only receive flow; they will not contribute to any downstream flow. A cell is considered to have an undefined flow direction if its value in the flow direction raster is anything other than 1, 2, 4, 8, 16, 32, 64, or 128.

In the graphic above, the top left image shows the direction of travel from each cell and the top right the number of cells that flow into each cell. Cells with a flow accumulation of zero are local topographic highs and may be used to identify ridges.

The flow accumulation map explains the direction of flow and its relationship with nature of the surface. The western portion of the study area is mostly covered by sloppy surface and this result into less accumulation of water falling in this region. The central part, especially on the location of Willington reservoir and the surrounding areas where the Manimukta Nadi, Gomuki Nadi, Sweta Nadi, Chinnar River and Anavari Odai merges with Vellar River. This illustrates the real condition of the surface over that area and also nature of the accumulation and its capability to hold the water over this area.

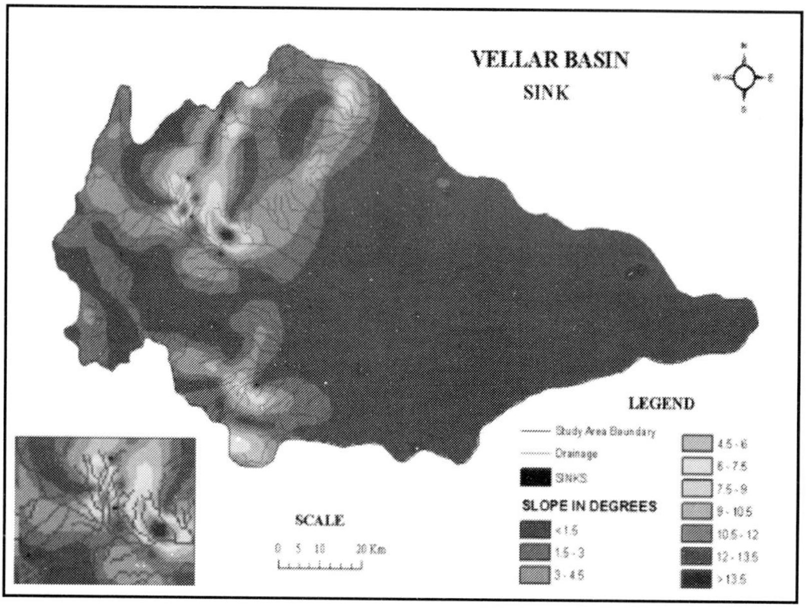

Fig 7. Sink

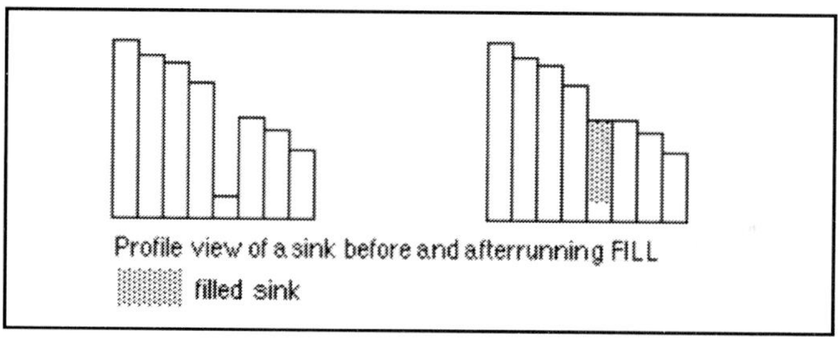

Fig 8.

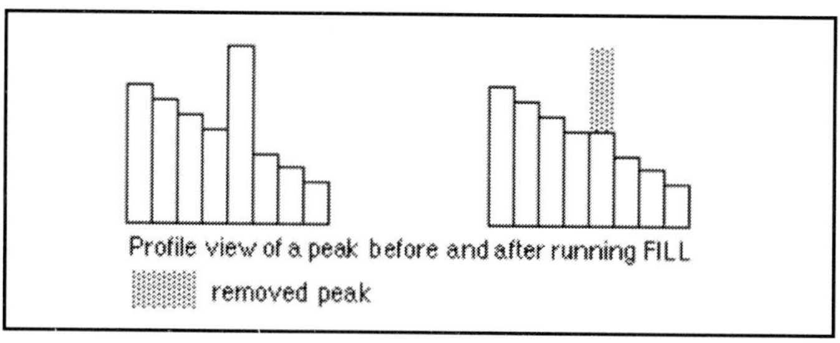

Fig 9.

Flow Direction

The Flow Direction function creates a raster of flow direction from each cell to its steepest down slope neighbor. The direction of flow is determined by finding the direction of steepest descent from each cell. This is calculated as: Change in z value / distance * 100. The output of Flow Direction is an integer raster whose values range from 1 to 255. If the direction of steepest drop were to the left of the current processing cell, its flow direction would be coded as 16. The flow direction function used to explore the nature of the river basin as per the topography. The direction is based on cell values in the raster grid (elevation). The direction flow will be based on center value of the kernel comparing with adjacent cells. In the Vellar basin the flow direction is towards east from western portion of the study area. Here the resultant map shows the directions as per the gradation of colours. The dark blue represents the northeastern direction (value of 1 in Kernel window), light blue represents the eastern direction (value of 128 in Kernel window). The main river Vellar passes through light tone, which represents east and southeastern direction in the study area.

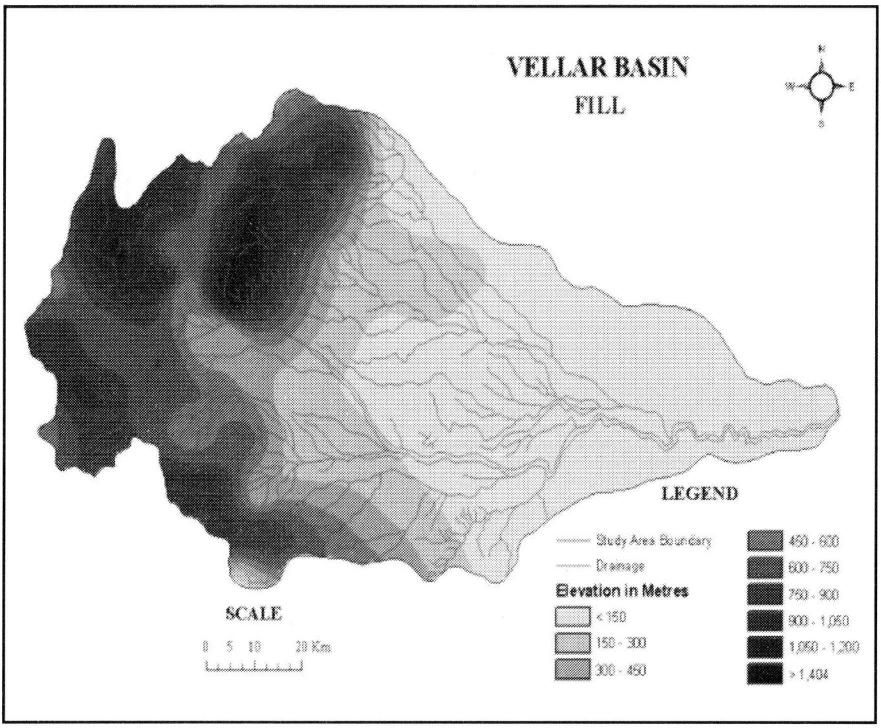

Fig 10. Fill

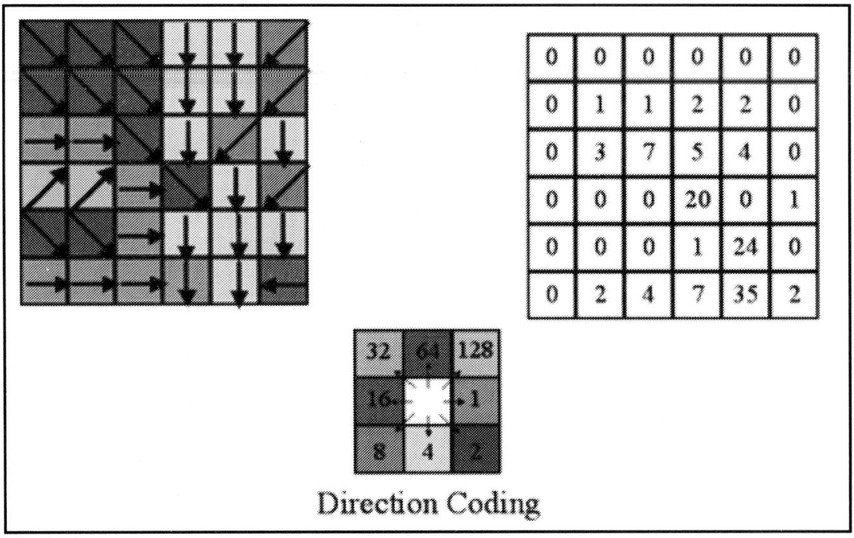

Fig 11. Direction Coding

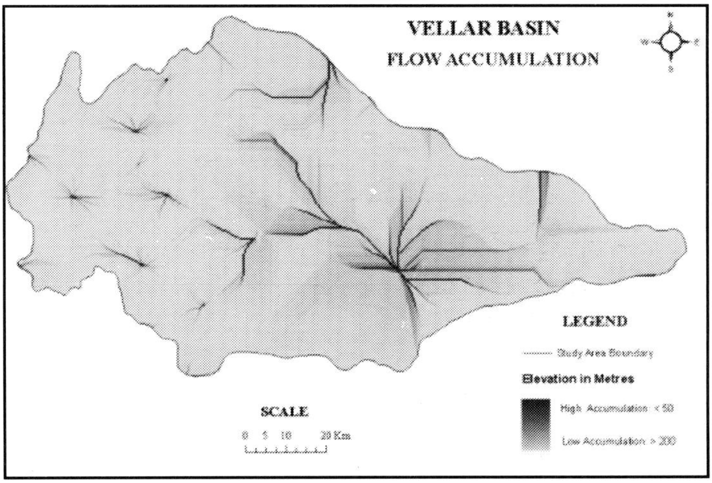

Fig 12. Direction Coding

Flow Length

A primary use of the Flow Length command is to calculate the length of the longest flow path within a given basin. This measure is often used to calculate the time of concentration of a basin. This would be done using the UPSTREAM option. Flow Length can also be used to create distance-area diagrams of hypothetical rainfall/runoff events using the weight raster as impedance to movement down slope.

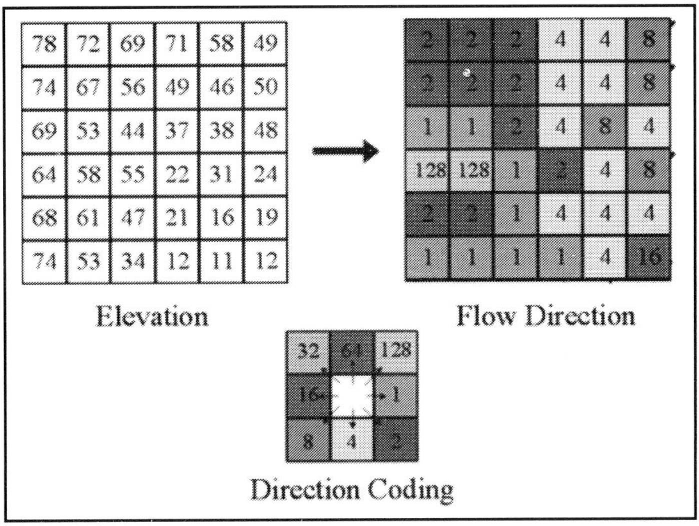

Fig 13. Flow Direction

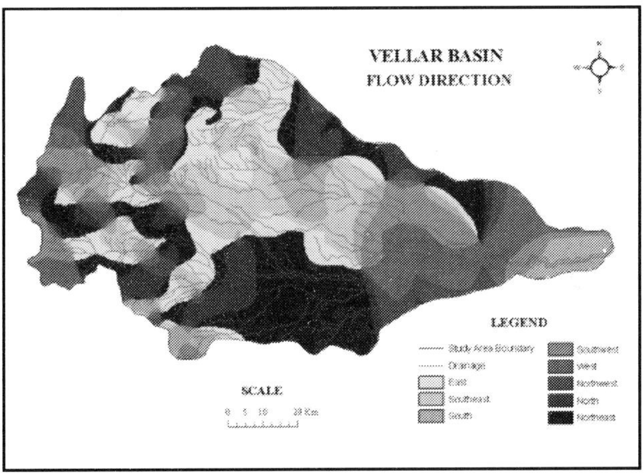

Fig 14. Flow Direction

DOWNSTREAM calculates the down slope distance, along the flow path, from each cell to a sink or outlet on the edge of the raster. The different colour gradation explains the down stream flow length as per given class interval. Here the class intervals are not equal because of distance of flow path may vary as per the topography of the basin. The central portion of the sub basin shows the gradual decrease of length. Here we can create distance - area diagram of hypothetical rainfall / runoff event using the weighted raster. Another use of this function is that we can find the length of each stream order. The western hilly region of the study area has shortened length of flow path and the long flow path is found in the middle and the eastern portion of the study area.

UPSTREAM calculates the longest upslope distance along the flow path, from each cell to the top of the drainage divide. The resultant map shows the upstream flow length is in descending order, as the less length has been found in the elevated terrains in the western portion of the study area by noticing the dark patches of gradation. In middle portion of the basin the streams have considerable length because of the plain surface and it is shown in light tone. The main function of this flow length tool is to calculate the upslope distance along the flow path, for each cell of the basin area to top of the drainage divides. Here the distance is measured in metres. This function is useful to calculate the travel time of water through a watershed.

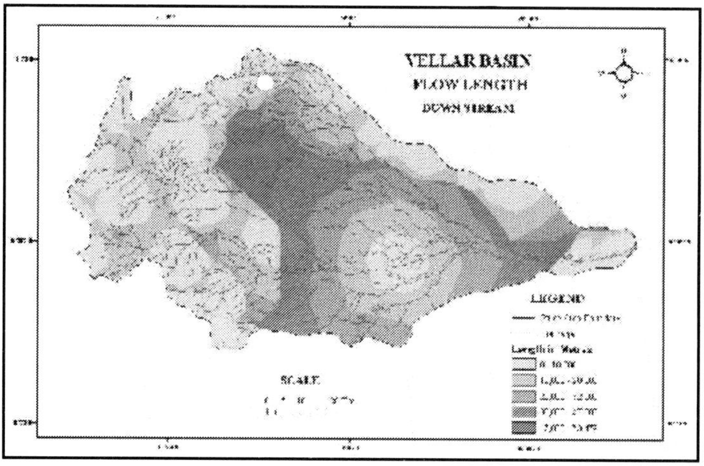

Fig 15. Flow Length - Down Stream

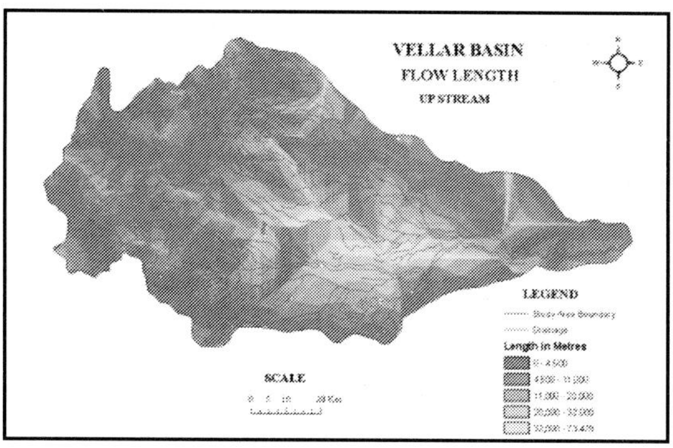

Fig 16. Flow Length - Up Stream

The main purpose of this tool is to show the points of highest accumulation of flow. In the study area the maximum number snap pour points has been found in the northwestern western portion and this may be due to hilly terrain. The less number of snap pour points has appeared in the central and eastern regions. The snap pour points in the elevated surface are shown in dark black spots and in the plain surface with white spots.

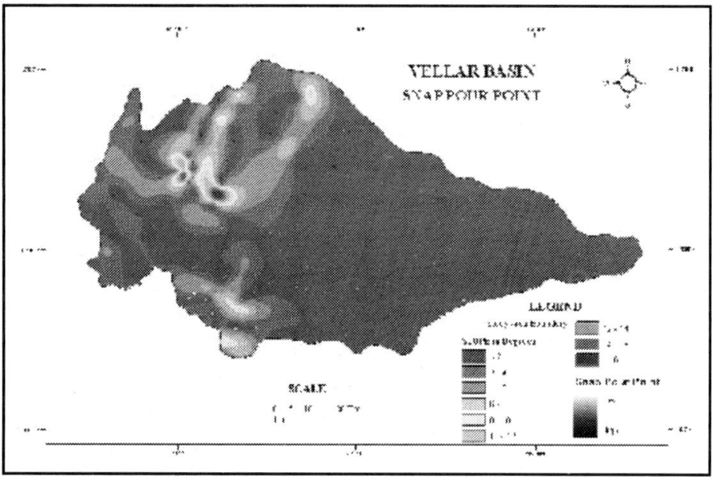

Fig 17. Snap Pour Point

Watershed

A watershed is an area that drains water and other substances to a common outlet as concentrated drainage. Other common terms for a watershed are basin, catchment, or contributing area. This area is normally defined as the total area flowing to a given outlet, or pour point. These areas are the output of the Watershed function. The boundary between two watersheds is referred to as a watershed boundary or drainage divide.

An outlet, or pour point, is the point at which water flows out of an area. This is the lowest point along the boundary of the watershed. The cells in the source raster are used as pour points above which the contributing area is determined. Source cells may be features such as dams or stream gauges, for which one can determine characteristics of the contributing area.

The entire Vellar basin has been taken as a single watershed area and the foothill areas of Kalrayan, Kolli and Pachamalai hills are water catchment areas of this basin. The Vellar basin is the fourth largest

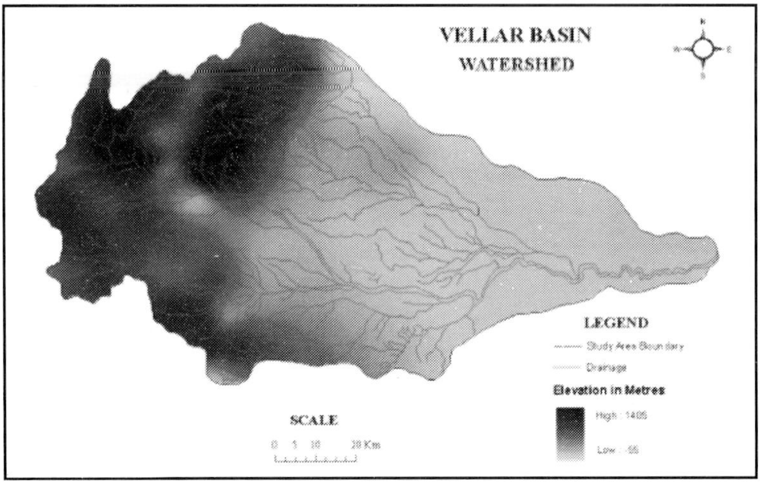

Fig 18. Watershed

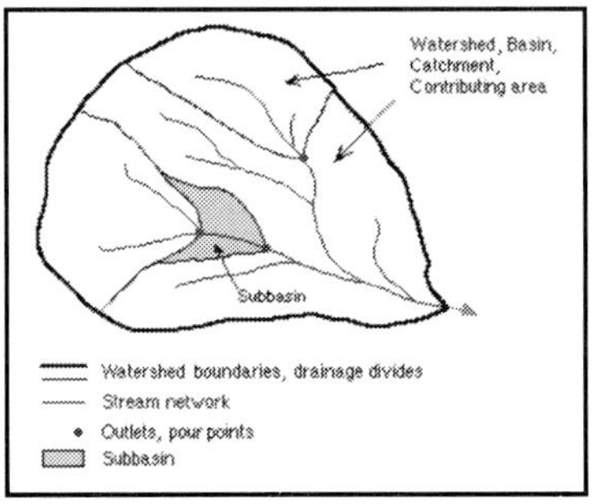

Fig 19. Sub - basins

basin in the Tamil Nadu state. The entire basin has around seventy-eight sub-basins as per the given elevation details. The outlet portion is the lowest point of elevation of this entire watershed area.

Summary And Conclusions

The aim of present study is to investigate the hydrological conditions of the Vellar basin, in Tamil Nadu using GIS. The applications and aspects of hydrological tools in ArcGIS have been explored and the hydrological information has been derived after investigation.

Understanding the hydrologic properties and functions of watershed structure and topology is central to understanding and predicting several important watershed parameters including stream flow and physical structure. Measurements of watershed metrics necessary to quantify, analyze, compare and contrast are often expensive to measure in the field and the large quantity of data required for coarse scale analysis is frequently unavailable. Utilizing GIS in conjunction with digital elevation models provides an innovative method to generate some of the requisite metrics necessary to investigate hydrologic properties in relation to watershed structure and geographical arrangement. GIS integrates all kinds of information and applications with a geographic component into one manageable system.

Hydrological analysis has been carried out with the help following layers such as drainage, slope, aspect and Digital Elevation Model. In the analysis part the Vellar basin has an area of about 7584 sq. km is divided into 78 sub-basins based on the input layers of flow direction and elevation. Sinks are considered to have undefined flow directions. By using flow direction as the input raster layer the sink map have been developed. The hilly tracts in the western portion of the Vellar region have been characterized with several sinks. The disadvantage of sink can be over come by the fill option. The Fill function fills the sinks in a surface raster to remove small imperfections in the data. The fill will require DEM layer as an input to develop the sink model.

The flow direction has been derived based on elevation surface. The resultant surface shows the normal flow of drainage from west to east. The flow accumulation map is generated by using flow direction layer as a input and the result shows the central portion of the Vellar basin as the higher accumulation of water than the remaining areas. The flow length map has been developed with the help of flow direction layer, as the input the direction of measurements can be estimated for both down stream and upstream.

The snap pour points function needs the input layers of Digital Elevation Model, pour point feature and accumulation raster layer. The output shows the snap pour points along the river course with different colors, which indicates the higher surface points and lower surface points. A watershed is an area that drains water and other substances to a common outlet as concentrated drainage. The input layers required are flow direction and feature pour point data. The output shows various catchment areas of this basin near to foothills of mountainous regions in the western portion of the study area.

Recommendations for further study:

- Water resource assessment for this study area.
- Estimation of hydrological effects resulting from changes in the physical characteristics of a basin
- Real time forecast using Meteorological forecast.
- Estimation of probable maximum flood from rainfall
- Assessment of ground water recharge, and
- Assessment of water quality.

REFERENCES

Analysis for Watershed Assessment http://www.rsmas.miami.edu

Brown. S., and Dee D.D, (1995). A case study - Using A GIS- Based Solution for Watershed Analysis and Automation. http://www.columbia.edu

Heywood, I., Corneliu, Steve, (2003). An Introduction to Geographical Information Systems. Singapore: Pearson Edition.

http://atlas.gc.ca/site/english/maps/environment/hydrology/drainagebasins

http://edcdaac.usgs.gov/gtopo30/hydro

http://geog.gmu.edu/PEOPLE/fall2004/GIS_653_4_Continuous_interpolation_2005.pdf

http://unescodelhi.nic.in/natural/hydrology.htm

http://www.epa.state.oh.us/dsw/gis/learngis/7append/glossary.html#flowdir

http://www.geog.ubc.ca/courses/klink/gis.notes/ncgia/u40.html#SEC40.1

http://www.gisdevelopment.net/technology/gis/ma03100abs.htm

http://www.ucowr.siu.edu/hydro/h04.html

John, L.G., (2003). GIS for Water Resources and Watershed Management. London: Taylor and Francis.

Kang-tsung, C., (2003). Introduction to Geographic Information Systems, New Delhi: Tata McGraw-Hill Publications.

Seth S.M., Jain S.K., and Jain M.K., (1993). Remote Sensing and GIS Application Studies at National Institute of Hydrology, National Institute of Hydrology, Roorkee.

Thornbury, D.W., (2002). Principles of Geomorphology. New Delhi: New Age International (P) Limited.

APPLICATIONS OF GEOSPATIAL TECHNOLOGY
A. Ganesh (Ed.), (2006) : pp. 151-157
Satish Serial Publishing House, New Delhi
E-mail: ssp@ndf.vsnl.net.in, hkjain1975@yahoo.com
website: www.satishserial.co

9

Creating Topology for Surface Water Features

P. SUNDARARAJ, R. NARAYANAKUMAR,
K. KUMARASWAMY AND K. SASIKALA

Topology is a branch of mathematics that studies the properties of geometric figures that are twisted, stretched, shrunk or otherwise destroyed without broken into pieces. In other words topology is geometry on rubber sheet. It is the particular field of geometry that is concerned with spatial relationship rather than rigid coordinates. The concept of topological relationship can be best explained by means of a topological map. In essence, a topological map contains explicit topological information on top of the geometric information expressed in coordinates (Masry and Lee, 1988).

In a topological map, all spatial entities are decomposed and represented in the forms of three basic graphical elements, viz., point, line and polygon. A unique feature identifier identifies each point entity. Each line entity (or arc) is identified by the line with its nodes. Polygon entities are formed by using the line entities and their respective nodes forms polygon entities. Once formed, polygon entities are individual entities identified by a unique identification number. The topological relationship among the polygon entities are computed and stored along with the line entities.

In GIS, topology is a model used to describe how features share geometry, and it is also the mechanism for establishing and maintaining topological relationships between features. In order to create the topology there are validation rules, which includes deleting duplicate objects, erasing short objects, breaking crossing objects, extending undershoots and snapping clustered nodes and a set of editing tools that work with features that share geometry in an integrated fashion.

Topological Relationships between Spatial Objects

The relationship in spatial data should be used logically with more complicated data structures. The following are some of the important topological relationships exist between spatial objects:

- Polygon to polygon relationship
- Polygon to line relationship
- Line to polygon relationship
- Line to line relationship
- Point to point relationship
- Point to line relationship

Significance of Topology

Topology applies behaviors to spatial data. Topology enables GIS software to answer questions such as adjacency, connectivity, proximity and coincidence. Topology provides a powerful and flexible way for users to specify the rules for establishing and maintaining the quality and integrity of spatial data. Topology can also be used to validate the spatial relationships between different features in a drawing or feature classes. The following are the functional areas of GIS where topological relationships are applied:

Spatial Search by Topological Relationship

The use of topological relationships makes a GIS perform spatial searches efficiently. It is a tedious process, to find manually the adjacency of various polygons. But in GIS the computer makes use of the adjacency information to identify immediately all polygon that share a common boundary with the parcel of interest and describes data from the geographic base. Similarly, connectivity information is used for spatial search using line data. The commonly used GIS function to find optimum path between two points is using arc-node relationships stored in the geographic database.

Construction of Complex Spatial Objects

Complex spatial objects are represented as complex polygons in geographic database. There are two types of complex polygon: those containing one or more holes also referred to as island or enclaves and those made up of two or more polygon that is not physically connected. Complex polygon can be easily visualized when displayed on the screen or printed on a map.

Types of Topology

There are two important types of topology used in this study, viz., link-node topology or network topology and polygon topology.

Network Topology

Network topology defines the interconnection of links and, optionally, nodes at link junctions. Links can be lines, open polylines, or arcs. Network segments have specific direction. Generally network topologies are used for direction analysis. The directions derived from line objects might be arbitrary. To control and edit these directions network topology information is stored on the links and nodes as object data. Each element of the network topology has different object data values. The Table 1 shows an object data table of a typical network topology.

Table.1. Object Data Table of Network Topology

Topology Name	Object Data Table	Object Data Field
Sample_Net	Tpmlink_Sample_Net	Id Start_Node End_Node Direction Direct_Resistance Reverse_Resistance

Polygon Topology

Polygon topology is an extension of network topology and focuses on area-based relationships. Every area forms a polygon and each polygon in a topology consists of a set of links. A polygon in a topology has a centroid, which is a point or block element within the polygon and contains information about the area it encloses. The following are the key points regarding polygon topology:

a) Polygons almost always share links; most links belong to two polygons, one on the left and one on the right,

b) Polygons can have 'islands;' the most common example is an actual island within a lake. Note that islands can also have islands,

c)Polygons have areas and perimeters, which are calculated and stored with the centroids and

d) Multiple topologies can share edges. For example, a river can be part of several topologies.

Before creating a polygon topology the data must be cleaned up. There should be no gaps, intersections, or overlaps between any of the line work in a polygon topology. Polygon topology can be created from closed polylines that share an edge or intersection with other polygons. Creating polygon topology is similar to creating network topology, except that in polygon topology centroids are created.

Any selected point, text, or inserted object inside a polygon is used for the centroid. Each centroid holds data about the number of links in the polygon and the area and perimeter of that polygon only. Links and nodes hold object data for all topologies that use them. Information about a polygon topology is stored in the centroids and links. Each centroid (Table 1) and links contains (Table 2) the following object data information:

Table.2. Object Data Table Attached to Centroids

Topology Name	Object Data Table	Object Data Field
Sample_Poly	Tpmcntr_Sample_Poly	Id Area Perimeter NKLinks_Qty

Table.3. Object Data Table Attached to Links

Topology Name	Object Data Table	Object Data Field
Sample_Poly	Tpmlink_Sample_Poly Left_Polygon	Id Start_Node End_Node Direction Direct_Resistance Reverse_Resistance Right_Polygon

Uses of Topology

Creating topology to the data in GIS has a variety of applications, some of which include:

i) *Cadastre:* Automated land parcel management integrated with property, transactional and financial data.

ii) *Boundaries and Population Management:* Managing people depends on counting and classifying them. This is done on an area basis.

Managing the boundaries of these areas is the prime application of topology.

iii) *Utilities, Network Management and Emergency Management:* Managing utilities such as roads, gas pipelines and electric cables are easy with topological relationships for e.g. shortest path identification is efficient only with topological relation parts. Topology enables accurate and valuable network routing, along with complex queries so that, in case of a fire, for example, explicit information can be delivered not only about the buildings affected but also about adjacent properties.

Study Area

The Pambar river basin is located at southern Tamil Nadu extend from 9°45′ to 10°25′ N latitudes and 78°10′ to79°0′ E longitudes. This river flows from west to east originated at Sirumalai of Dindigul district, covering Dindigul, Sivagangai and some parts of Madurai, Pudukkottai and Ramanathapuram districts. The river joins at Bay of Bengal near Sundarapandiyanpattanam. The main tributaries are Manimuttar, Devakottaiyar and Varshali. The length of the river is about 150 km form the place of its origin to the point of confluence in the Bay of Bengal.

Methodology

The study is carried out with the help of two software AutoCAD and ArcGIS. Survey of India topographic sheets 58 J/3, J/4, J/7, J/8, J/11, J/12, J/16, 58 K/9, K/9, K/13, 58 N/4, 58 O/1 and O/2 in the scale of 1:50,000 are used for digitizing the tanks and drainage of Pambar river basin. The following is the step by step sequence of creating topology for Pambar river basin. The steps one to seven are carried out using AutoCAD software, while steps eight to twelve are carried out using ArcGIS software.

1. Geo referencing the topographic sheet.
2. Digitizing the map
3. Cleaning the drawing
4. Object selection, object conversion, clean up option
5. Extend undershoot and trim the overshoot
6. Creating topology
7. Expor ting to ArcMap

8. Cre ating coverage
9. Finding the area, perimeter and length
10. Calculating total Area
11. Report
12. Layout

Results and Discussions

The study area has around 1350 tanks. One third of the river basin is filled with surface water bodies like tanks. The agriculture, domestic and other water usages heavily depend on the tanks of the river basin. The basin contains a large number of smaller tanks having less than 5 km^2 area. The major tanks which are more than 10 km^2 in area could be located around the central and lower southeastern portions of the basin. The number of tanks is in negligible amount on the catchments especially on the western side of the river basin. The density of drainage is high on the western side of the catchment, where Sirumalai is located. The density of the drainage progressively decreases towards the central and southeastern side of the basin, which is the ayacut area of the basin.

The area, perimeter and adjacency of the tanks and the length connectivity of the streams are calculated by creating topology using AutoCAD for the study area. The object data table contains all the above said information. Through this rigorous exercise of topology, the Pambar river basin has been made suitable for an accurate GIS study. From this study the basin area is topologically clean without under shoots and over shoots. Each and every tanks and streams are tagged with id. The basin is spatially locked (geo-referenced). The tanks and streams are measurable in terms of area, perimeter and length and ready for any type of spatial query.

REFERENCES

Albert, C. P. L., and Yeung, K.W., (2002). Concept and Techniques of Geographical Information Systems. New Delhi: Prentice-Hall of India.

Chang K.S., (2002). Introduction to Geographic Information Systems. New Delhi: Mc Graw-Hill.

Masry, S.E., and Lee, Y.C., (1988). An Introduction to Digital Mapping. Fredericton, New Bruns Wick: Department of Surveying Engineering, University of New Brunswick.

Heywood, I., Cornelius S. and Caver.S, (1998). An Introduction to Geographical Information System, Delhi, Pearson Education.

http://www.dcs.st-and ac/uk.htm.

http://www.esri.com/new/arcnew/sammer02article/arcgis.bring.topologt

http://www.geo.ed.ar.uk/agridexe/term.

http://www.innovativegis.com/basis/primer/concepts.html.

http://www.mala.bc.ca/com/

http://www.ordnance survey. co.uk.

http://websis.wr.usgs.gov/" globalsis/metada-gr.htm.

APPLICATIONS OF GEOSPATIAL TECHNOLOGY
A. Ganesh (Ed.), (2006) : pp. 159-167
Satish Serial Publishing House, New Delhi
E-mail: ssp@ndf.vsnl.net.in, hkjain1975@yahoo.com
website: www.satishserial.co

10

Application of GIS for Delineation of Functional Zones in an Urban Area

G. R. PARTHASARATHY AND RAJU GOPINATH

The later half of the 20th Century witnessed a significant shift in the process of world urbanization which made the developing countries to come in the limelight. Many of the developing countries have a large base population and this has led to the large scale addition of absolute number of urban people though the percentage increase is very less. Provision of amenities to such a rapidly growing urban population has become a Herculean task. Further, many of the conceptual models that have been developed based on Western cities could not be applied to the cities in the developing countries without modification. Logically this has initiated a spate of urban research articles analyzing the various dimensions of urbanization process in the developing countries.

Technically, the whole subject of Urban Geography can be summarized as the "Study of city 'in' area and city 'as' area" [Carter, 1972] based on the scale dimension. In small scale, cities become points in a region and the analysis mainly focuses on the relative importance of these points among the given set of points. Fundamental to this analysis is the 'functions' or 'services' offered by the urban centres. The number and status of services bring varied dimensions of the urban centres formulating distinctive sets of systems of cities with hierarchical relationship. Identification of spheres of influence also falls under this purview.

On the other hand, in large scale, the city itself covers an area and the intra city dimensions form the core theme of analysis. Outlining this theme of morphology or internal structure of the city, Carter [1972] illustrates that the analysis could be done taking either architecture or layout or land use as the fundamental aspect. While architecture or layout has been taken as the basis for urban research by other disciplines, urban geographers have been heavily dependent upon land use as their major tool of urban research. The present attempt has also followed the same geographical approach of taking land use as the basic criterion for analyzing the morphology of a metropolitan area.

Morphology of the City

Division of a city into morphological zones has been attempted even as early as 1920s [Burgess 1923]. This has initiated a number of other conceptual models [Hoyt, 1939; Harris and Ullman, 1945] including Social Area Analysis [Shevky, Bell and Williams, 1955] and Factorial Ecology [1965]. Though these models have identified different zonations within the city, only CBD has been considered for detailed analysis. A number of attempts have also been made to delimit CBD.

William and Olsson [1940] selected 'shop rent index' to delimit CBD. Volume of sales has also been taken as the basis to include an area within the CBD. However Murphy and Vance considered several variabies such as population density, land use, land value, rent and distribution of selected institutions and tried to use them for delineation of CBD. They developed four different indices namely, Total Height Index [THI], Central Business Height Index [CBHI], Central Business Intensity Index [CBII] and Central Business Index [CBI]. Basic aspects considered in these indices are the area of floor space used for central business and the total area of the building.

In countries like the USA block wise data are available in the census and hence delineation could be done easily. In the case of developing countries such types of data is not available. Further the occurrence of mixed land use also complicates the delineation. In addition many Indian cities are old and possess different morphological characteristics than that of the western cities. Given this situation, the application of Geographical Information System has become a powerful tool and helps to solve some of the problems of delineation of morphological zones within a city. Taking Thiruvananthapuram as the study area, an attempt has been made to use GIS for morphological zone delineation.

Study Area

Thiruvananthapuram is the head quarters of Kerala State. It is located between 8° 26′ and 8° 32′ N latitudes and 76°53′ and 77° 01′ E longitudes. It covers an area of 74.93 km^2. It had a population of 744739 as per 2001 Census. As one of the oldest cities in the region it had its origin as settlement as early as 8th Century A.D. However the real growth as an urban centre came only after British period. The first areal survey was conducted in 1876 and the area identified was only 5.9 km^2. The town was divided into five sectors for administrative purposes. It was declared as a municipality in 1920 and a corporation in 1940. After the formation of Kerala state in 1956 it became the administrative capital. The area of the city increased to 74.93 km^2 in 1962 and remained the same till 2005.

With elevations varying between 0 and 75 metres above sea level the city has a general slope from northeast to southwest. More than 55 per cent of the city area comes under side slopes with slopes varying between 5 and 18°. Karamana and Killiar rivers form the major drainage features and there are a number of smaller channels. The city enjoys a warm tropical maritime climate with an annual rainfall of 180 cm. An analysis of the growth rate of population indicates that except 1961-1971 decade the city has only a low to moderate growth of population. Population density varies between 35 persons per hectare to more than 420 persons. In conformity with most of the Class I cities of India, the population density is very high in and around the city core area. Similarly it also follows the distance decay principle with densities falling when one moves towards the periphery. With such a background the city exhibits varying patterns of land use distributions. For an effective analysis of the land use pattern and change in varying periods the city has to be divided into various morphological zones. For this the packages used are PC ARC\INFO 3.5 and ILWIS 2.2.

Methodology

The analogue data in the form of maps, air photos, etc are first scanned and stored as tiff files. Since this format is not compatible with GIS, it was brought to AutoCAD platform to perform on screen digitization. Before digitizing equal sets of control points have been fixed. Thematic maps such as ward maps, road maps, land use maps, etc for different time points have been scanned and digitized. After performing error checks and editing the maps were exported in DXF format. They have been then exported to ARC\INFO as 'coverages' and stored in workspace. Using ARCEDIT module the final topology was built with BUILD command with necessary creations of Polygon Attribute Table [PAT] and Arc Attribute Table [AAT]. Since the derived maps were drawn on different projections, they have been re-projected to Polyconic projection using control points. Appropriate tabular data has been created with Look up table or (LUT) and Feature Attribute Table and they were also merged using a relational join. Thus each thematic layer in the data base has been encoded, corrected and transformed to create an integrated GIS data base ready for analysis.

Delineation of Morphological Zones

To delineate the morphological zones of the Thiruvananthapuram city, the latest point of the time has been considered. The units of observation used are

1. Spatial pattern of the land use
2. Land value surface

The land use and land values are the symbiotic dynamic variables, which operate in the urban system. The spatial structure of these variables is closely related to distance from the city centre, amenities and population density. Therefore, a change in any of these variables simultaneously induces change in other related attributes. Hence, these variables, with the same intensity can be isolated into compartment to facilitate judicious interventions. This interfused association of these variables makes them the ideal units for morphological zonations.

Spatial Pattern of Land Use

For the purpose of zonations, the categories of land use of the city have been simplified into residential and non-residential land use. The residential class covers the area occupied by residential houses. It also includes garden lands, paddy fields, wetlands and other undeveloped areas of the city which can be converted to residential land use. The non-residential class envisages all the other built up areas occupied by commercial, industrial, administrative and other institutional establishments. Further, with this specialized class, land use coverage, ZLUCOV (Figure 1) has been prepared for this period. Simultaneously ward coverage, WDCOV, has also been prepared of this period. Further, WDCOV has been overlaid over ZLUCOV, to form a new coverage LUWDOCOV. The tool used for the overlay of coverages is 'UNION', which computes the geometrical intersection of two polygons

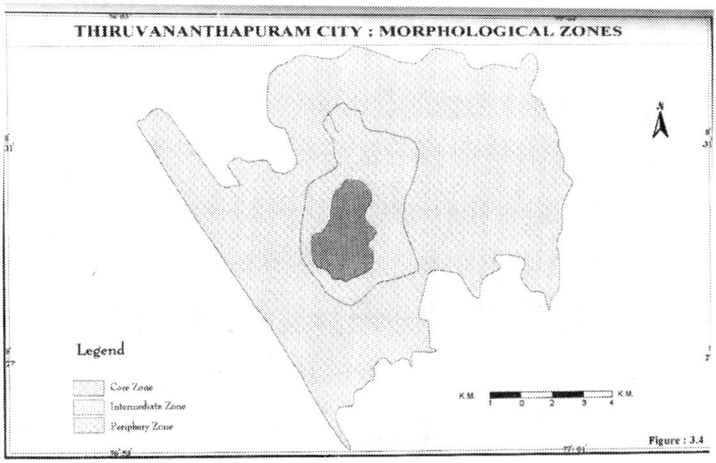

Fig.1. Thiruvananthapuram city: Morphological Zones.

in which all the features and attributes of both coverages are preserved. From LUWDOCOV, the percentage of area under non-residential land use to the total land area of each ward has been estimated. This calculated percentile score of non-residential land uses in each ward has been marked correspondingly. Further, depending on the intensity of variation in score, the values have been grouped into high (above 80%), medium (between 80% and 60%) and low (below 60%). The areas of wards falling in each group has been differentially marked (Figure 2) using differential identification land use (lucode) codes.

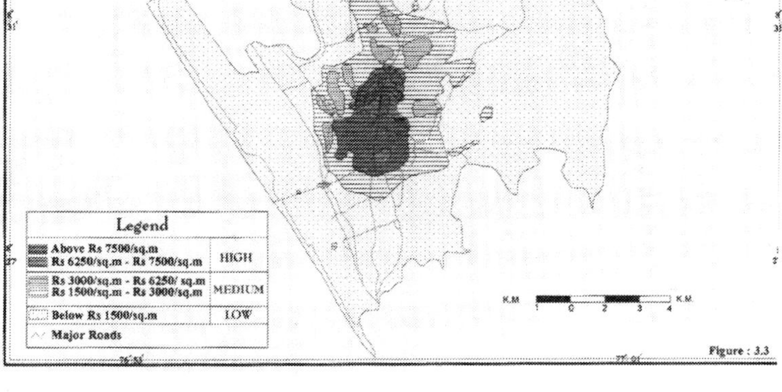

Fig.2. Land Value Surface for Zonation

For the demarcation, emphasis is given on non-residential land use. However, a few exceptions are also made. For example, an airport is usually located in the outer area. Therefore, in a calculation, that area may be having a higher value. Similar is the case of jail, military area and heavy industry. It such cases, though they possess high values, they are not included for delineation and areas are shown as unassigned areas in the map.

Land Value Surface

The spatial distribution of land value in the city has been observed from the revenue records and through field surveys. From these observations, it has been noticed that heterogeneity of land value surface follows the transportation lines and, to a limited extent, the land uses. So for tracing these land value figures spatially, road coverage, ROADCOV, has been selected. This coverage consists of boundary, roads and water bodies of the city. The nature of land value distribution

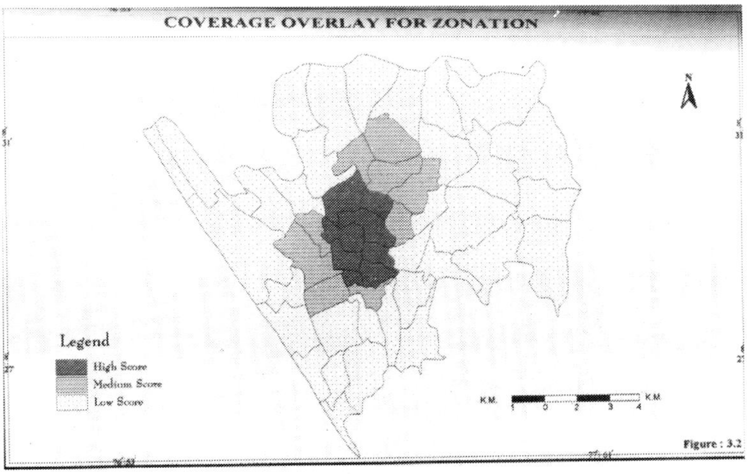

Fig.3. Coverage overlay for Zonations

in the city has been thoroughly observed and the values have been grouped into high (above Rs 6250/m^2), medium (between Rs 6250/m^2 to Rs 1500/m^2) and low (below Rs 1500/m^2), depending on the variations. From the observations, it has been noticed that highest land values follow the main arteries and nodes. The medium land values follow the sub arteries and the areas lying immediate to the high land values. The remaining areas have low land values. For tracing these trends of land value variations on to the ROADCOV, conditional buffering technique has been employed. For buffering, the conditions applied are the same as that of those observed from the field survey. Thus the spatial distribution of the land values has been successfully marked on the ROADCOV to form a new coverage, LVRDCOV. Further, polygons formed by each land value group have been differentially marked by differential identification land value (lvcode) codes (Figure 3).

Integration and Finalization

For either unit of observations, i.e. for land use as well as land values, two different coverages with different identification codes, i.e. lucode, have been prepared. To facilitate zonations, the LVRDOCOV has been overlaid using 'UNION' over LUWDOCOV to form a new coverage, LULVOCOV that has the polygon features and attributes of both the coverages. The demarcation of the zone boundary has been done by the conditional reselection of polygons to form a new coverage LULVRES in which map features from a coverage (LULVOCOV) based on feature attribute values has been extracted.

The conditions set for reselection are as:-

i. For boundary separating first zone from second zone: Percentile share of Non-residential land use=80 per cent and land value=Rs. 6250/m ^{2}.

ii. For boundary separating second zone from third zone: Percentile share of Non-residential Land use =60 per cent and land value=Rs 1500/m ^{2}.

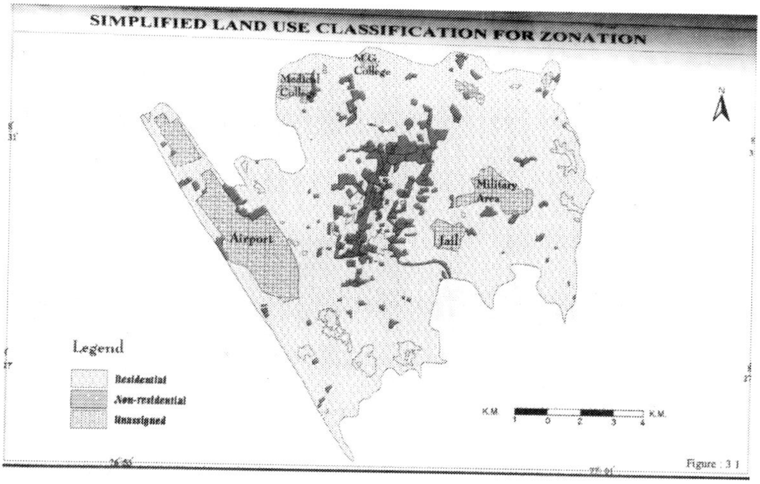

Fig.4. Simplified Land UK classification for Zonation

Thus three morphological zones with precisely demarcated boundaries have been identified as shown in the map (Figure 4). The inner zone has been named as the core zone and the outer zone as periphery zone. The zone lying between them is named as the intermediate zone. It is not out of place to give a brief description about the general characteristics of these three regions.

Core Zone

This is the central built up zone of the city, which occupies only 5 per cent of the total city area. The actual areal extent of the zone is about 373 ha. It is the principal center of administrative and commercial functions and hence the residential functions are comparatively less significant. The percentage of non-residential land use to total land use of the zone exceeds 80 per cent in 1998, while it was only 50 per cent in 1981 and 40 per cent in 1966. This zone records the all time highest land values of over Rs 6250/m^2 in 1998, but had a value of only Rs 1250/m^2 in 1981 and Rs125/m^2 in 1966.

Intermediate Zone

The zone encompasses the core in an irregular fashion with more spatial extension in the eastern and northern sides. It covers about 13.5 per cent of the total city area and can be marked as the functional offshoot of the core. The residential landscape admixtures the non-residential land use and hence the percentage of non-residential land use and land values is moderate. The percentage of non-residential land use to total land use of the zone varies between 60-80 per cent in 1998, while it varies between 25-50 per cent in 1981 and between 20-40 per cent in 1966. The land value varies between Rs 6250/sq.m to Rs. 1500/m^2 in 1998, Rs 1250/sq.m to Rs. 125/m^2 in 1981, Rs. 125/sq.m to Rs. 25/sq.m. in 1966.

Periphery Zone

This is the outermost zone of the city where the residential land use outstrips the non- residential land use. This zone occupies the major share (about 82 per cent or 6109.8 ha) of the city area. The percentage of non-residential land use to total land use of the zone is below 60 per cent in 1998, while it was less than 25 per cent in 1981 and 20 per cent in 1966. This zone also records low land values, which can be attributed to the distance decay effect. The land values are below Rs. 1500/m^2 in 1998, Rs. 125/m^2 in 1981 and Rs. 25/m^2 in 1966. Thus the morphological zonation has been done with the help of GIS tool. It may be seen that this is only an arbitrary method since the values chosen for delineation have been derived only based on subjective appraisal of the overall pattern. Application of similar methodologies in other case studies may help to develop a model for delimiting morphological zonation using GIS.

REFERENCES

Carter, H., (1972). The Study of Urban Geography, London: Arnold.

Gopinath, R., (2001). Spatio-Temporal Analysis of Urban Morphology of Thiruvananthapuram City Using Geographical Information Systems. Unpublished Ph.D. Thesis. Mysore: University of Mysore.

Mayer, H., and Kohn, (1966). Readings in Urban Geography. New Delhi: Prentice-Hall.

APPLICATIONS OF GEOSPATIAL TECHNOLOGY
A. Ganesh (Ed.), (2006) : pp. 169-175
Satish Serial Publishing House, New Delhi
E-mail: ssp@ndf.vsnl.net.in, hkjain1975@yahoo.com
website: www.satishserial.co

11

Land Use Change Detection

G. R. PARTHASARATHY AND RAJU GOPINATH

Land is a finite resource. Though water is always called the elixir of life, land becomes more important for humanity because man is a terrestrial animal. In fact it is our habitat and has been shared by other living organisms. After 1650 the population explosion that has taken place in the world has resulted in higher concentration of human population on this fragile resource. Further not all land areas are suitable for human life. In fact more than 60 per cent of the land area is either too cold, or too dry, or too wet, or too rugged so that comfortable human life is not possible. Though man is also living in these so called 'Negative Areas' or 'Non-Ecumene' his number is too small. Thus the ever increasing population is assaulting more and more on only about one-third of the land area. The last hundred years saw a surge of scientific and technological advancements which helped man to modify natural environment to suit his short term needs. In many cases man has lost sight of the need for environmental conservation and he has paid heavy price for that also. A welcome change that has come in the late eighties of the last century created an opportunity for humanity to correct its wrong deeds by coming together as one society in preserving our Earth to make it livable.

Given this background the purpose for which land is occupied by man has also shown considerable modifications. The magnitude of these land use changes varies between one country to the other. Further even within the country they show various dimensions. A notable aspect of the world population is that it is being shifted towards urban areas. At the beginning of 20th century only 10 per cent of the world population was living in urban areas. By 2004 this has increased to nearly 48 per cent. It is expected that after 2007, there will be more urban population than the rural population in the world. In conformity with this shift there will be considerable change in the pattern of land use in the urban areas. Since most of this growth of urban population is concentrated on the developing countries, cities in the developing countries are undergoing dynamic land use changes. Realizing the importance of this the International Geographical Union has indicated Land Use / Land Cover Change Detection [LUCC] as a major research theme.

Analysis of land use changes of a metropolitan city provides considerable challenges. The non-availability of data and cost make it difficult for conducting such studies in the developing countries. Further cities in the developing countries vary from their counterpart in the developed countries in their growth processes. Sources of data for land use varies from topographic sheets, city maps, aerial photographs and satellite images. Integration of these sources into a single frame has posed much problem for geographers. The development of Geographical Information System has helped to solve this to a considerable extent. Taking Thiruvananthapuram, the head quarters of Kerala State the present study tries to analyze the land use changes between 1966 to 1998 using GIS.

For the present study of land use changes different sources of data were collected. Maps from the Town Planning Department, Survey of India toposheets, aerial photographs and IRS IC images are used to generate thematic maps for selected time points. It should be noted that for the first time point the primary source of data is the town planning department map and the SOI toposheet. An intermediate time point of 1981 is also taken to know more about the trend of the changes. For the latest time point satellite imagery is used. Necessary field checks have also been made for ground truth verification. Analogue data thus created in the form of thematic maps have been brought into GIS format. The usual steps of scanning, digitizing, correcting and map re-projecting were done. Necessary control points have been fixed on the thematic layers for different time points so that overlay option becomes easier. The software used for the analysis and mapping are AUTOCAD, PC ARCINFO and ILWIS 2.2.

Status of Land Use in 1966

Figure 1 shows the distribution of different types of land use in 1966. It may be seen that nearly two-thirds of the city area was under residential land use. It is in conformity with the pattern that any Class I city in India has shown. This land use pattern is also distributed throughout the city area. Additional thematic layers have been created for this time point taking the distribution of population density and the residential household density. The effect of distance decay has clearly been visible in the east, north and northeast. Only in the southwest a minor deviation could be noticed. This is due the presence of Poonthura slum which has increased the density in southwest.

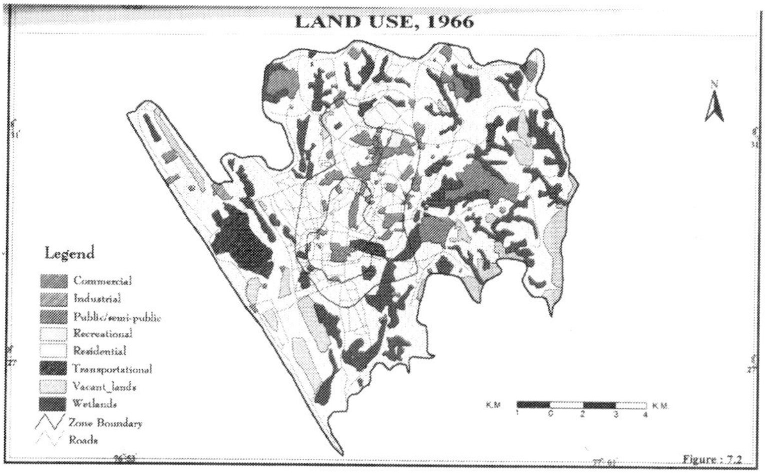

Fig.1 Land Use 1966

There is a considerable area under public and semi-public land use category which accounted for about 7.7 per cent of the total area of the city. This is not surprising since Thiruvananthapuram is the head quarters. This type of land use, however, occurs more as distinct patches in the central, northern and eastern parts. The largest of these patches is in the east comprising of the Military area and the Central jail. Medical college in the northeast and a group of buildings including the Secretariat along the MG road in the north central part are the other notable areas of this type of land use. In addition the Sree Padmanabhasamy temple in the core area also stands out as a distinct patch. Airport, bus station and railway junction constitute the major areas for transportation land use. While airport which accounts for more than two-thirds of the area under this type of land use lies in the west, the bus station and railway junction occur as one patch and are found in the central area of the city.

Commercial land use has only a share of just 0.4 per cent. It occurs mostly along MG road and in Palayam area. Industrial and recreational land use types also cover some area [Table 1]. In the case of undeveloped parts of the city the wetlands are more significant and they cover nearly 1100 hectares of land. This occurs as patches in the periphery regions especially in the eastern part. Most of the patches in the east are low-lying paddy fields which follow Killiar and Karamana rivers while the patches in the west are low-lying marshy waterlogged areas. In addition to wetlands, some vacant lands are also found particularly in the southeastern and southwestern peripheral regions of the city.

Table.1 Percentage Distribution of Various Types of Land Use in Thiruvananthapuram City

Land Use Type	1966	1981	1998
Residential	67.02	74.48	77.50
Commercial	0.35	0.76	1.67
Public/Semi-Public	7.73	9.53	10.11
Industrial	0.24	1.30	1.56
Transportational	2.96	3.22	6.19
Recreational	0.28	0.62	0.78
Wetlands	14.50	10.09	2.19
Vacant land	6.91	0.00	0.00

Status of Land Use in 1981

The pattern and distribution of land use in 1981 has not shown much change from 1966 [Figure 2]. However, some rearrangements have been noted as given in Table 1. It may be seen that vacant land has been completely removed and converted to other uses. Residential land use continues to be the most dominant type of land use. It has increased its share to nearly 75 per cent of the total area of the city. Nearly 560 hectares of land have been additionally brought under this land use between 1966 and 1981. Much of this is from vacant lands and wetlands particularly in the peripheral parts of the city. The population distribution layer also brings out the fact that the high density areas are gradually pushed out from the core towards the periphery. Conversion to residential land use is particularly higher in Karamana and Manacaud areas. Public and Semi-Public land use continued its increase during this period also. Its share now is nearly 10 per cent of the total area of the city. While the distribution pattern remains almost the same as that of 1966, some new patches have come up in the central part of the city particularly in Palayam area.

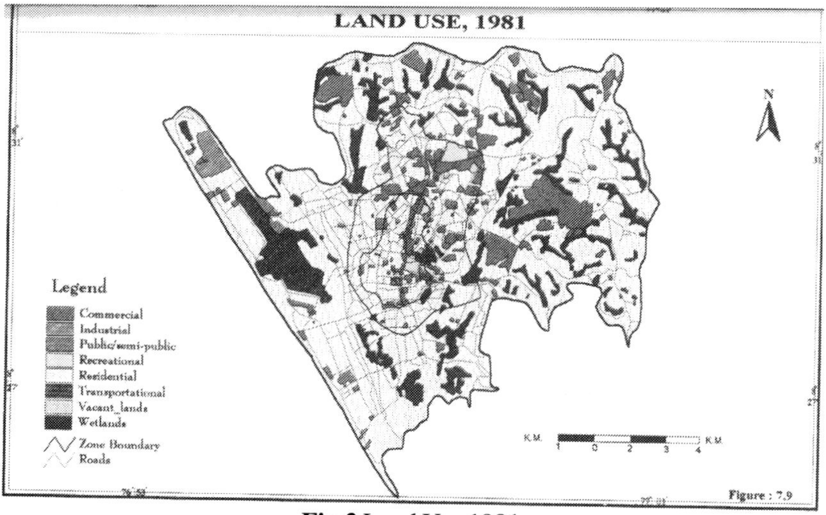

Fig.2 Land Use 1981

Considerable increase is noted in the case of area under transportational land use mainly because of expansion of airport land. Similarly industrial land use also has increased its share particularly in the western areas. A small industrial patch has been developed in the northwest part of the city. In addition Hindustan latex limited, which is established in the northern part of the city also added its share.

Commercial land use has shown only marginal increase. New patches of commercial areas have sprung up in the Chalai-East Fort-Karamana belt and in the Palayam area. The Putharikandom maidan located near the Sree Padmanabasamy temple is now used for recreational purposes and this is mainly responsible for a marginal increase of this land use in the city in 1981. Wetlands have shown a sizable decline between 1966 and 1981 while the vacant land has been totally eliminated.

Status of Land Use in 1998

Figure 3 shows the distribution of different types of land use in 1998. It may be noted that there has been considerable rearrangement of the land use pattern between 1981 and 1998. Residential land use continued its supremacy during this time point also. But the increase in the share is comparatively less. Increase in the density of residential houses is particularly noticeable in the eastern and northern periphery of the city. Public and Semi-Public land use has also undergone only a slight increase in its share between 1981 and 1998.

The declaration of Thiruvananthapuram airport as the Fifth International Airport in India has brought considerable changes in the transportational land use of the city. This resulted in nearly doubling of the area under this type of land use. New commercial areas have also come up in Vazhuthacaud—Karamana areas in the central part of the city as well as in the northern peripheral areas like Medical College area, Kesavadasapuram and Peroorkada area. The Veli industrial belt developed in the northwestern part is mainly responsible for increase in area under industrial land use. A marginal increase in the recreational land use is noted mainly due to the inclusion of Sankumukhom beach as well as the development of backwater tourism in Veli area.

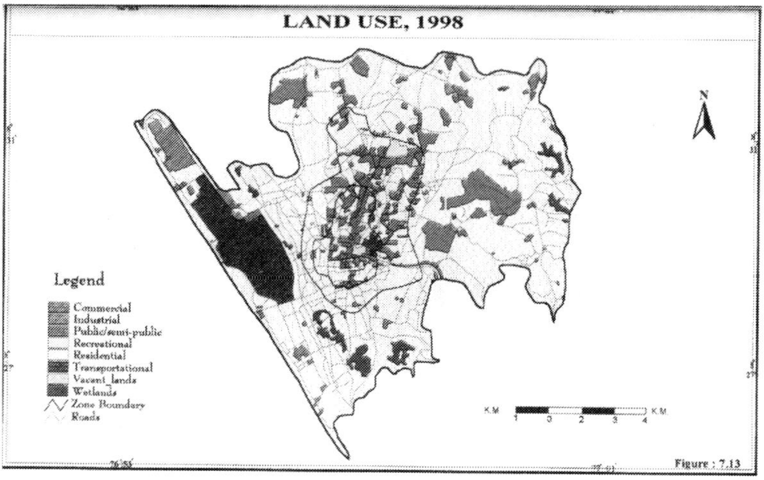

Fig.3 Land Use 1998

Changes in Land Use between 1966 and 1998

The salient features of land use changes during the two end time points in Thiruvananthapuram city are:

- Area under residential land use has increased from nearly 67 per cent to 77 per cent between 1966 and 1981 bringing out the fact that still the urban area has not truly converted itself as a major metropolitan area.
- The administrative function of the city has resulted in the distribution of public and semi-public land use as a distinctive north-south band of area in the central parts of the city.
- A nearly six-fold increase is noted in the commercial land use but the share is considerably lower compared to the cities of similar size in other parts of the country.

- Transportational land use doubled its area mainly due to elevation of airport as an International Airport.
- Industrial areas have also shown a nearly six-fold increase. A well defined industrial belt is being developed in the northwestern part of the city which may increase more industries in future.
- All the vacant land has been converted to other land use particularly to residential use.
- A drastic decline in the wetlands could be noticed. This land use type has lost more than 900 hectares to various types of urban land use during the study period,

REFERENCES

Carter, H., (1972). The Study of Urban Geography. London: Arnold.

Mayer, H., and Kohn, 1966, Readings in Urban Geography New Delhi: Prentice-Hall.

Raju Gopinath, (2001). Spatio-Temporal Analysis of Urban Morphology of Thiruvananthapuram City Using Geographical Information Systems, Unpublished Ph.D. Thesis, Mysore: University of Mysore.

APPLICATIONS OF GEOSPATIAL TECHNOLOGY
A. Ganesh (Ed.), (2006) : pp. 177-183
Satish Serial Publishing House, New Delhi
E-mail: ssp@ndf.vsnl.net.in, hkjain1975@yahoo.com
website: www.satishserial.co

12

GIS Applications in Market and Competitive Intelligence

P. S. VIJAYALAKSHMI

Spatial tools are increasingly used common in IT Infra Structure especially in corporate data bases. Geographic Information System is a tool kit for creating spatially-enabled information systems and for managing and analyzing business information of any kind according to where it is located. GIS helps to manage customers, manage vehicle fleets, locate new businesses, create and manage marketing campaigns, organize sales territories, analyze retail spending patterns, provide better services, and understand the competition for our products etc. GIS extends the study of information systems by including spatial data, spatial processing and spatially mediated knowledge. In this paper an attempt has been made to study how GIS helps in improving performance of marketing communications.

Intelligence is a term that is increasingly applied to a range of external business functions. It may be used synonymously with information or news. It may describe an actual business process. Intelligence as a formal discipline began to take shape in 1980's. Since the beginning of commerce and competition collecting business information and deriving insights and taking decisions were in practice. Increasing competition, globalization and technology were beginning to dramatically alter traditional business practices, perspectives and stakes. Hence business sought ways to harness increasing amounts of data and information about customers, competitors and industry. Government intelligence models and practices were being introduced and adopted to the commercial sector. Arising out of these developments, competitive intelligence is a key factor in the current business environment.

Applications of Geospatial Intelligence in Marketing

Characteristics Predictor

The location of an individual, a family or a business play an important role in predicting the characteristics of that individual a family or business. For analyzing information from household data bases or from census statistics a detailed picture of the demographic features of an area or location can be developed. This area profile has been proved to be very accurate in predicting the characteristics of the make up of those areas. The important point to note is that the area demographic profile is built on the basis of where those individuals or families are so it is through where they are; we begin to know something of who they are. Understanding of customer characteristics and behavior helps to improve the vendor/customer relationship and performance of the transaction to mutual benefit.

Determining and Predicting Past, Current and Future Behavior

As a development from the prediction of neighborhood characteristics, analysis has also been used to predict the potential or future behavior of an area. Using the methodologies developed to predict character of an area, behavioral predictions can be made to peer into the future likelihood of an area to behave in a certain way. Central to this calculation is the location in question and the characteristics of this location. Behavioral predictions and understanding of character of locations have been extensively utilized by location specialists for store placement and real estate development.

Timing and Mechanics of Action - Extension of Behavior Prediction

Location can also play an important role in the timing and mechanics of the purchase decision, for example, the weather patterns dictating at what point purchases becomes likely. In addition to this, there are the mechanics of the purchase that can be influenced by location; highly urbanized areas see high levels of delivery services used.

As an extension of timing the purchase the actual products purchased is closely related to location, weather patterns, proximity to store, product range carried by location. Different physical locations need to stock a different product range to ensure appropriateness to the local conditions and local population. Natural calamities such as tsunami, storms such as Katrina, Rita and earth quakes, its frequent occurrence and location intelligence can form market opportunities for insurance and risk taking companies and health care institutions. Such type of market intelligence can be developed by using spatial techniques or GIS.

In the multi-channel world at present the understanding of a customer's preference and likelihood to respond via a specific channel, like telephone or mail, is crucial. Location plays an important role in not only predicting behavior, but in determining the most practical means of utilizing channels to suit a consumers needs based on where they live.

Extrapolation of the Known to the Unknown

Geographic location is the key factor in being able to extrapolate known information onto persons, households or areas that up to that point had little or no information. Through the extrapolation of information we are able to know, or more accurately infer, intelligence that can be crucial to business growth and organizational performance. It is through this extrapolation of information that location becomes one of the most valuable, yet under utilized, tools available to business and organizations.

MI and CI Market intelligence (MI) and competitive intelligence (CI) are complementary and can overlap in analyzing customers and markets. For example, CI and MI may each conduct or use positioning research to better understand how customers view products, services and brands compared to the competition. MI is concerned with understanding markets and customers in order to support decisions relating to market opportunities, new market development and market penetration strategies. To accomplish this, MI research and analysis tends to be based on market research techniques, including focus group research, segmentation research, and price elasticity testing and demand estimation. CI, while considering customers and markets, tends to treat these as two of a number of forces that impact competitiveness and decision-making, and rarely focuses primarily on these factors.

Competitive information gathering (or competitive research) is the component of CI that leads up to analysis, but lacks meaning and insights. Often, firms that simply conduct competitive information gathering mistake this practice for true CI. Organizations may gather news and other information about competitors or their industry - even generate informational products like alerts or newsletters. Without applying formal or informal analysis, as well as aligning the activity with a decision need, however, this practice stops short of intelligence. Having defined competitive intelligence, how is CI practiced?

The fields of Competitive Intelligence and Location Intelligence are inter-related. Here is a diagram explains that inter-relationship:

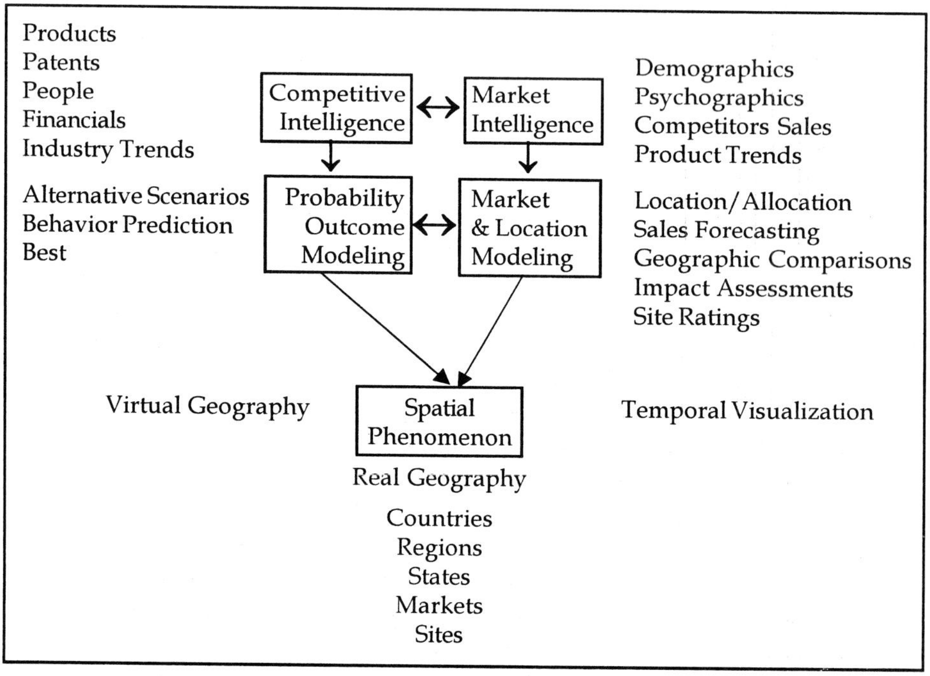

Because of the inter-relationship, there are opportunities to use the techniques of each to ultimately assimilate "Spatial phenomena" that drive an enterprise knowledge base into corporate policy.

Intelligence Functions

CI should be established according to the specific intelligence needs of the organization and its decision-makers. Good CI and useful insights rely on good intelligence questions, quality information, appropriate analytical tools and experienced practitioners. Intelligence Cycle defines specific functions for establishing a full-service intelligence unit. It begins by identifying the intelligence needs of the intelligence users (i.e. decision-makers). These needs are typically defined as key intelligence topics (KITs) which generate key intelligence questions (KIQs). KIQs are the specific questions on which intelligence research will be conducted. For example, a KIT may involve a competitor's plan to establish a new manufacturing plant closer to a particular market. Some of the KIQs may include:

- What is the timeline for location a site, building the plant, establishing full manufacturing capabilities/capacity?
- What regulatory issues may they face?
- What is the cost of establishing the facility? Operating costs?
- Which products will be manufactured?
- Who will be the key suppliers?

In addition to intelligence needs, other user needs should be defined: deliverables timeline, ways the intelligence will be used, format of delivery, etc. once the needs and questions are outlined and understood, project planning takes these into account. The project manager outlines a plan of action, resources, tools and personnel. This includes identifying analytical models, mapping information sources and establishing a timeline for each function and for the final delivery.

The research function for intelligence involves two types of resources: published sources (also called secondary research or literature research) and human sources (also called primary research). Each is highly specialized in intelligence gathering and relies on skills specific to its function. In published source collection, a professional searcher or corporate information center trained in intelligence and business research gathers content from print or digital material. These materials range from competitor's brochures to government filings to news articles to photographs. Human source collection derives information and expert insights from people knowledgeable about the topic or issue. This function is performed by skilled interviewers (often former journalists, industry specialists, and trained researchers), who may speak with former employees of a competitor, industry analysts, suppliers, customers, etc.

The analysis function is often essential to generating the intelligence and insights that are the aim of CI practice, particularly when there are multi-part KIQs, complex questions or a large volume of data/information involved. Intelligence analysis can involve more than 100 different models, spanning business functions, types of operation and industries. This function relies on intelligence information that may be collected using published sources, human sources or both. Taking into account the KITs/KIQs at hand, quantitative or qualitative models may be used.

Delivery is vital to the successful use of intelligence. Intelligence deliverables should be distributed to the right intelligence users, address their intelligence needs, be timely and be in usable form. These may range from scheduled reports and bulletins to products that address ad hoc intelligence issues. It is important to point out that news alerts may be vital information sources, but they are not substitutes for intelligence deliverables, which are tied to particular intelligence needs and often include analysis and recommendations.

For CI the Intelligence Process should be considered a framework and a starting point. Each function - and the process itself - requires customization to the specific needs and characteristics of an organization. Moreover, organizations may apply CI by developing selected functions internally and outsourcing other functions, as needed. For example, businesses often conduct routine published source collection on their own, but may outsource highly specialized research (e.g. patent or trademark searches), formal interviews of human sources or complex analysis.

CI and LI

The relationship between CI and the emerging practices and tools that comprise location intelligence (LI) is rich and mutually beneficial. First, the significance of CI to LI can be found in the well defined and well established process, practices and applications of competitive intelligence. The term "location intelligence" is often used to describe the tools involved with mapping and business network analysis. This emphasis on tools (and data) over process and techniques may limit the capacity to fully generate intelligence. The generation of location intelligence may be more targeted, efficient and effective by modeling CI's use of formal frameworks, its business discipline and its focus on decision-support. This will encourage a consistent application of location intelligence and help ensure that LI fulfills the intelligence need of decision-makers.

Competitive intelligence may be used to gain insights on the location-based capabilities of a competitor, partner or supplier. In better understanding a company's use of RFID in managing their distribution chain, for example, formally applying CI can tie this effort to the decision needs of the intelligence users, enhance the quality of the intelligence and help generate more usable results.

Conclusions

CI itself can more consistently apply location-based tools, technology, and applications to gather information, perform analysis and generate insights on competitors, markets and industry. Some CI practitioners have begun to adopt LI resources and techniques, like mapping and visualizing collected data, applying business geographic, gathering information from more widely available satellite maps and applying location-based analysis techniques (e.g. retail network analysis and Spatial Temporal Computing). Cost and complexity are sometimes barriers, but increased awareness and education about LI will encourage more practitioners to use these tools and enhance the generation of intelligence.

REFERENCES

Gilad, B., (1990). Business Blindspots: Replacing Myths Beliefs Assumptions with Market Realities, Wilt shire: Infunortis, England.

Hal Reid, (2004). Competitive Intelligence and Location Technology, Directions Media Publications.

Hernandez, T., Erguden and P Bermingham, (2004). Market thresholds – Major Retail Chains in Canada. Ryerson: CSCA Ryerson University Publication.

Mark Smith, (2004). Market Share Growth through Location and Business Intelligence, CRM Project.

Tianming Hu and Sam Yuan Sung, (2003). Spatial Similarity Measures in Location Prediction, Journal of Geographic Information and Decision Analysis, 7 (20), pp. 93-104.

Yeates and W Cheng,(2002). North American Markets – Predicting Metropolitan Growth and Change. Ryerson: CSCA Ryerson University Publication.

APPLICATIONS OF GEOSPATIAL TECHNOLOGY
A. Ganesh (Ed.), (2006) : pp. 185-194
Satish Serial Publishing House, New Delhi
E-mail: ssp@ndf.vsnl.net.in, hkjain1975@yahoo.com
website: www.satishserial.co

13

Geospatial Technology for Monitoring of Development Plan

SULOCHANA SHEKHAR AND JAYMALA DIDDEE

Now-a-days, major cities are competing with each other for attracting international capital and other investments and Pune city is no different. It pressurises the planning authority and the administration to cope with the existing as well as future demands. A town planner who aims for a sustainable development, has to face many challenges like expectations of people for better civic services, environmental amelioration, transparency in governance and so on. A problem common to all but often related to major cities is that many elements of their infrastructure have remained unchanged from the time when first put in place and they require enormous resources to extend them or modify them. Railroads, bridges, sewers, water mains, major roads and buildings have not been able to keep up with the expansion of many cities because of the speed of that expansion and because of cost. Using conventional planning methods will not help to resolve these problems. There is an urgent need for new, more flexible and less expensive technologies, and for a new organization of the city. Planners and decision makers are required to introduce innovative measures and techniques in their planning and decision making processes. State-of-the-art technologies like remote sensing, GIS and GPS proved their efficiency time and again in various fields and are imperative for the urban planning. Further to meet the people's requirement and imbibe local wisdom in the action plans, the spatial information base is pre-requisite (Krishna Murthy, et. al., 2000). In this paper, RS, GIS and GPS technologies were applied in monitoring the land use pattern of Balewadi village as per the draft development plan and this attempt could be a starting point to introduce these technologies in monitoring the development plan proposals in a dynamic fringe areas.

Study Area

The study area Balewadi is located in the northwestern corner of Pune city. This small village is located at 73°48 E longitude and 18°34'N latitude, with an area of 4.48 km². It is part of Aundh, a rapidly growing area of Pune city. It is just few kilometers away from Hinjewadi IT Park, one of the promising features of Pune's growth. Adding to that, it is an entry points those who come from Mumbai since the expressway passes through it. It is bordered by Mula River on its north and northeast. The southern part is connected to Baner, which is metamorphosing into happening place. It holds lot of green, barren land hidden for its future growth (Fig. 1).

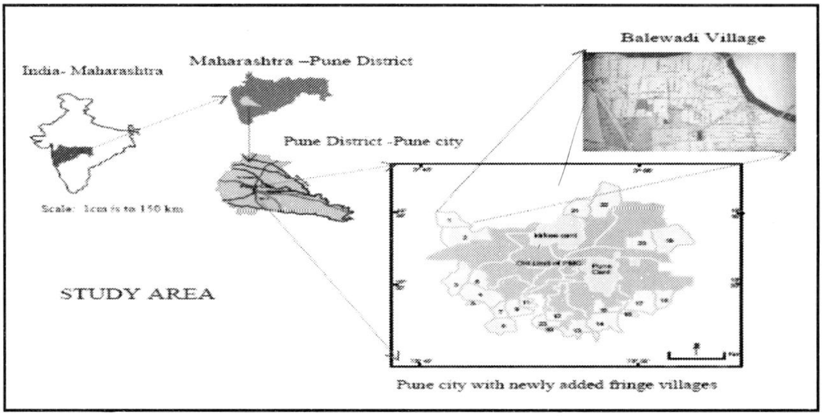

Fig.1 Pune City - Balewadi Village

In fact Pune's location was once considered less conducive for urban growth. But now Pune, the cultural capital of Maharashtra, is rapidly metamorphosing into a commercial hub. This metamorphism has a very long history spanning over1000 years or more. Pune, which has been hailed as 'Happening City', has not only survived but it has also prospered continuously during the last 400 years. Up to the mid seventeenth century, Pune was one of many small garrison towns on the plateau. Within a century, however, this fairly insignificant small town outstripped the older established towns in its respective region and emerged dominant. In the 19th and 20th centuries, it had undergone significant changes and prospered in various fields. Referred to as Oxford of East for its educational tradition and Detroit of India for its post independence industrial development, in the 21st century, Pune is emerging as a 'Cyber City'. The city is experiencing a high population growth of 62.17 per cent (Census, 2001) and it continues to grow. This increasing population of Pune city exercises a lot of pressure on the land available in the city and the city's limits have been pushed towards the adjacent villages. Recently, 23 villages had been added to the old municipal limit and the total area of Pune city increased from 145.92 km^2 to 243.96 km^2. This rapid growth needs to be monitored and properly planned for its sustainable development, particularly, the rural urban fringe areas require proper planning that will make the goal of enhancing the positive attributes and reducing the dysfunctionalities of a city.

Objective

The proper implementation of development plans and managing the urban fringe areas more sustainably are challenging and relevant issues of contemporary Indian urbanisation. Using geospatial technology as a tool, to explore the advantages and the efficiency of these technologies for proper monitoring and managing the development plan is the main purpose of this study.

Data Base

As the satellite images help to maintain truthful record of terrain during that period, IRS P6, LISS IV image was considered for this study. With the spatial resolution of 5.8 m, in multi spectral mode, LISS IV data is highly suitable for urban studies. IRS P6 LISS 4Mx. Image of 13th March 2004 (Scene No: 111) along with SOI toposheets of 47F/ 14 and 15 (1979-80), on 1:50,000 scale, Pune city Guide map on 1:20,000 scale and Draft Development Plan of Balewadi and Baner from Pune Municipal Corporation on 1: 10,000 scale were used for this study. The land use details are from the primary survey where as the demographic details were taken from census abstracts (Fig. 2a and b).

Methodology

The digital remote sensing data was processed and georeferenced in Erdas 8.3.1 software. Initially the toposheets were scanned and georeferenced and used as base for image registration along with GPS points. Using necessary enhancement techniques the georeferenced image was further enhanced. The Draft development Plan map was scanned and georeferenced with the remote sensing data so that

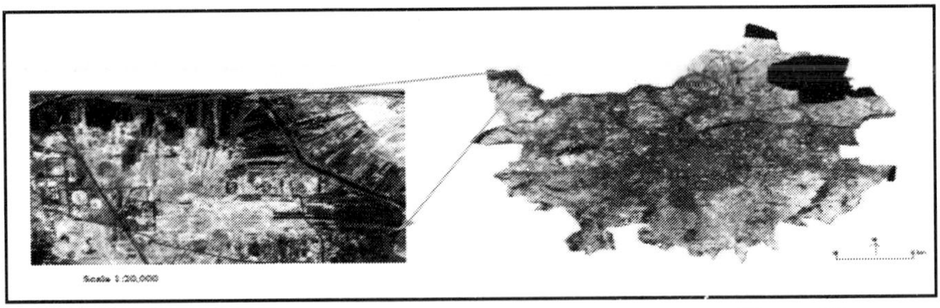

Fig.2(a) Balewadi in LISS IV image **(b)** LISS IV image of Pune City

overlaying of these two data made possible. Then land use classes were digitized from the draft plan by using Arc view. A detailed field survey was carried out at Balewadi village. The Garmin GPS was used to do the survey of all the building blocks (Residential /Commercial / IT industries). The block corners were taken as GPS points and the points were plotted on graph paper to get the lay out. After preparing the layout of building blocks, the layout map was scanned and brought into Erdas Imagine 8.4 for georeferencing. Then the georeferenced lay out map was transferred into (ArcView 3.2 a) GIS environment and the layouts were digitized into polygons for further analysis.

Development Plan

As per the Town Planning Act, it is mandatory for any city to prepare or revise the existing Development Plan for a future period of 10 or 20 years. The process of preparation of Development Plan in general involves a few steps. It starts with an analysis of spatial changes over a time period for a given region or city/ town. This is supplemented by study of demographic and socio economic changes for the corresponding years. Projection of population and employment together with estimation for provision of basic infrastructure, services and investment is the second step. Prescribing land use zoning to cater to the projected needs and in conformity with the set objectives and policies is the last step in formulating a draft plan. A draft plan is to go through an elaborate process of public consultation. The draft plan revised on that basis is to be sanctioned and notified by the government for coming into force. It is clear that the whole process is complex. It involves voluminous data analysis, examination of various development options and frequent presentation of information to variety of groups in both text and graphic forms. Some of these activities are to be completed in time bound manner too (Patkar and Sampath Kumar, 2001)

The old Bombay Town-Planning Act (1954) was replaced by the Maharashtra Regional Town Planning Act (MRTP) of 1966. This was in response to the unrestricted industrial growth in the Bombay-Pune region and the felt need to have a more integrated and coordinated development of the metropolitan region falling within the influence of Bombay. Under the MRTP Act, the state government had established a special area known as Pune Metropolitan Region (PMR), and set up a planning authority board for a new Development Plan. The Regional Plan for the Pune Metropolitan Region was sanctioned in 1971. At that time itself, it was recognized that some city-oriented development spilling over into the fringes beyond the Pune Municipal Corporation (PMC) limits was inevitable. Substantial housing development proceeded without proper planning and permission in the fringes of the PMC. This was by and large devoid of or largely lacking in facilities and amenities necessary for good urban living. It was, therefore, decided that the fringe villages should be brought into the ambit of the adjacent Municipal Corporations as these alone will be better able to provide the inputs necessary for improving the quality of life of existing residents, remove the current shortcomings of the un regulated development that has emerged, restructure and control future development on orderly lines. Accordingly, under notification in 1997, 38 fringe villages contiguous to the PMC limits were merged in the Pune Municipal Corporation. Meanwhile on 17 November 2001, the State Government revised its earlier notification and excluded fifteen villages fully and five villages partly from the PMC limits. At present only 23 fringe villages have been merged with in the PMC, in which 18 villages, were fully merged and five villages, were partly included. The City Improvement Committee (CIC) and General Body approved Development Plan (DP) for these 23 villages on 27th December 2002.

While finalizing the draft DP for 23 villages, PMC's planning committee has ruled that 1600 ha of hill land in those villages be reserved as Biodiversity Parks. Simultaneously construction in the 60 m distance from the riverbanks has been banned with the ultimate aim of Green Development Plan for Pune's 23 fringe villages. Achieving the aims and goals of green development plan for the fringe villages and sustainable development of Pune city now, depends on the proper implementation of these plans. It needs continuous monitoring and updating the existing plans and data. It is envisaged that high-resolution satellite images and GIS will serve various large-scale GIS operations in the areas of municipal GIS, town and country planning, facility management, disaster management and natural resources management (Kalyanaraman 2005). In this juncture, geospatial technologies play a critical role in regular monitoring and provides a complete information system to accomplish the sustainability. In order to prove the efficiency of this technology, a small case study was undertaken and the results obtained were worthy of emulation and could over the next few years change the face of Pune.

Results and Discussion

Actually the rural urban fringe is a space to create exciting new landscapes for the 21st century. Proper planning and implementing the planning proposals are therefore necessary to maintain the fringe areas more sustainably. The quality of the rural urban fringe is an indicator and advertisement for the quality of a town or city and its rural hinterland. As a gateway to the urban area, it creates a powerful first impression to visitors and the possible investors. Consequently the rural urban fringe especially along the major highways should be managed and maintained to a high standard. Considering this, a fringe village in the northwestern corner of Pune City, Balewadi was selected for case study. It acts as a gateway to Pune City and is just located enroute to Mumbai along Mumbai-Pune Express way. The selection of the study area was also based on the entropy analysis (Shekhar, 2003) where the entropy value of Balewadi village is quite high and it is also part of Aundh area, the choice destination of many people from Pune and Mumbai. (Shekhar, 2005).

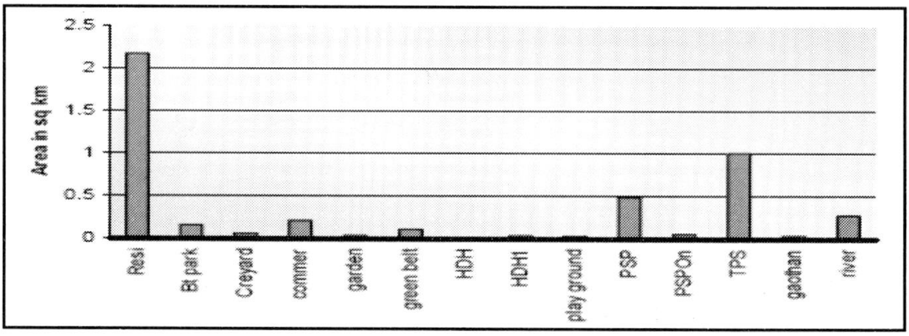

Fig.3 Proposed Development Plan of Balewadi Village

Balewadi's biggest advantage is its close proximity to Hinjewadi-IT park, (less than 5 km) and Part of Aundh, which is an upcoming area at a rapid manner. Being flanked on north as well as east by Mula River, it has a scenic location and lot of virgin land to its asset.

Therefore amble of chances to plan and implement the planed proposals to bring regulated growth and sustainable development (Fig.3). Out of the total area as per the DP, the residential should cover 48.7 per cent excluding the old Gaothan area. As per the survey, including Gaothan the total built-up (residential, commercial and public and semi public classes) area is only 11.81 per cent. Other interesting facts in the survey are only 8.1 per cent of the land was sold in Balewadi, for residential purpose (Fig. 4). Particularly, the sale was increased after the inclusion of this fringe village to the old PMC limit i.e., in 1997, when the 38 fringe villages were added to the old PMC limit, and in 2003, when the 23 fringe villages were finalized to add to the city limit. The survey numbers close to Pimpri-Chinchwad and Baner area, the sale percentage was quite high. For example in Survey No 11, the total land sold out was 92.93 per cent and in Survey No 22, the percentage of land sold out was 61.96 per cent. The rest of the survey numbers, the land sale was very less. In fact within the last five years (1997-2002), the land value showed an increasing trend. It shows there are more chances to develop the village as per the DP proposals

Resi: Residential Use; Commer: Commercial; PSBON: Octori Naka; BT Park: Biotechnology Park; HDH: High Density Housing; TPS: Town Panning Scheme; Creyard: Construction Material Yard; PSB: Public and Semi-Public Zone

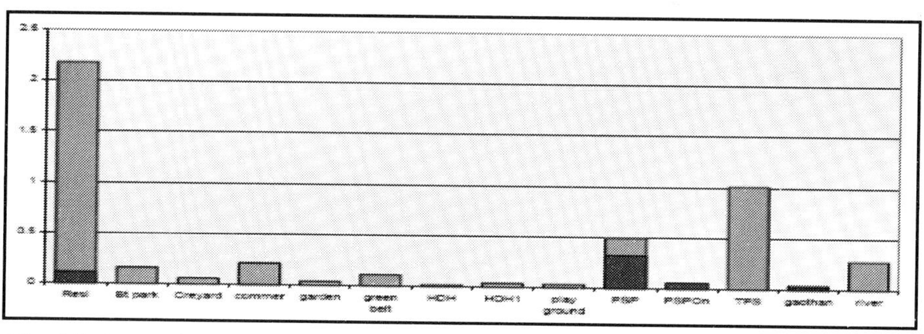

Fig.4 Existing land use Vs Proposed Development Plan of Balewadi Village

Proposed Land Use Classes in the Draft Development Plan of Balewadi Village
Existing Land Use Classes in the Village

The availability of high-resolution data is an added advantage to the town planners as well as municipal corporations. Now-a-days, many municipal corporations started using GIS as a tool in their planning process. For example Mirzapur, Gorakhpur municipal corporations used GIS to facilitate the municipal property assessment (Vaishali Nandan, 2004). In fact, GIS coupled with high-resolution data and GPS can be effectively used for monitoring and implementing the development plans. It can also be used as LIS (Land Information System). Thus, all information pertaining to a property, for example, land use, location, number of floors in the building, year of construction, year of stay of the resident, status of resident (owner/renter), rent paid (if the property is rented), type of construction, number of rooms in the house, area of rooms, total area and covered area of the property, was collected and stored. Infrastructure related information (availability of water, sewer and electricity connections) can also be collected and stored. Even attempts were made to sketch the plan of each property by using GPS, for future reference and comparison. The results from GPS survey confirmed the LISS IV image because the built-up area occupied only 11 per cent of the total area. The following figures show the series of steps followed in this analysis by using Arcview 3.2a (Fig 5 and 6).

When the draft plan proposals were digitized and overlaid on the LISS IV image, it becomes easy to identify the deviations of planned proposals (Fig. 7 and 8). Interestingly there were no deviations found and it gave us a hope to manage the land uses in a systematic manner as per our objectives. The layout map prepared by using GPS points, collected during the field survey was overlaid on draft DP and the layouts were exactly under the respective land use classes like, National Insurance Academy, NICMR, Bharathiya Vidya Bhavan English School, Octroi Naka, Sports City

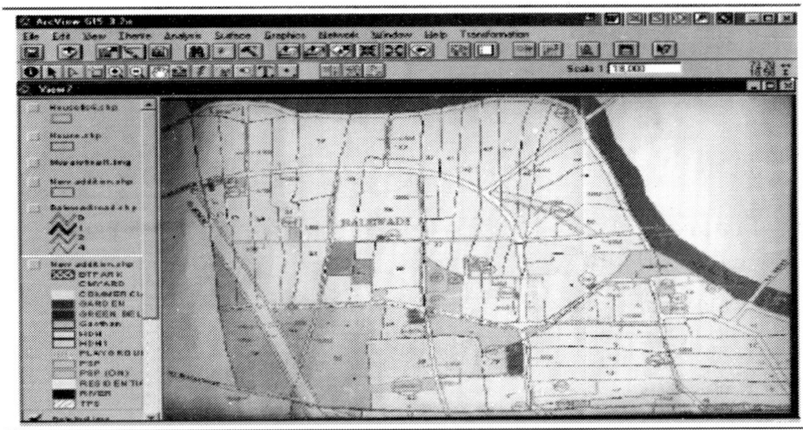

Fig.5 Draft Development Plan of Balewadi Village

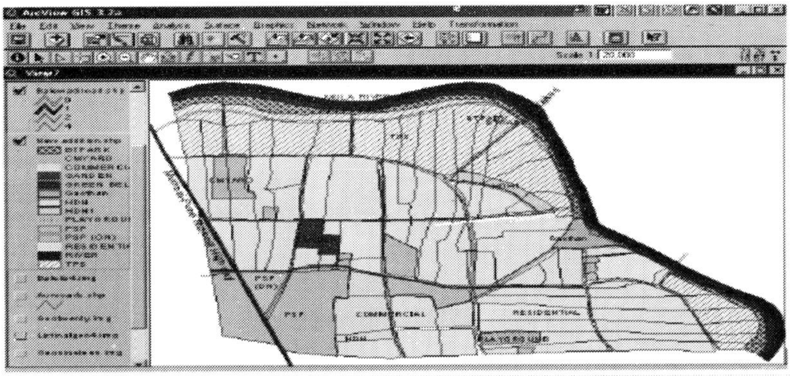

Fig.6 Draft Development Plan digitised in Arc view

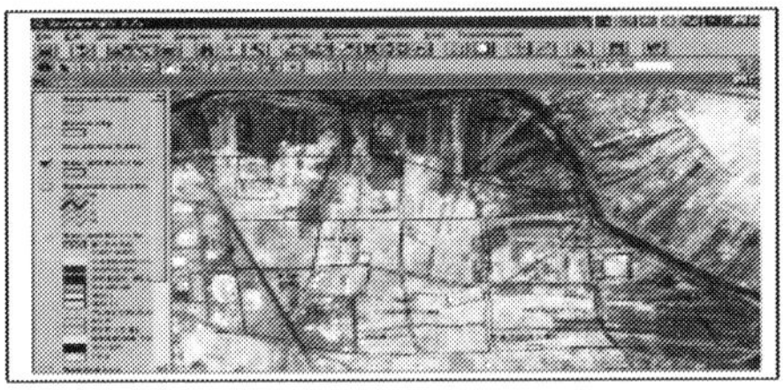

Fig.7 Draft Development on LISS IV image in Arc view

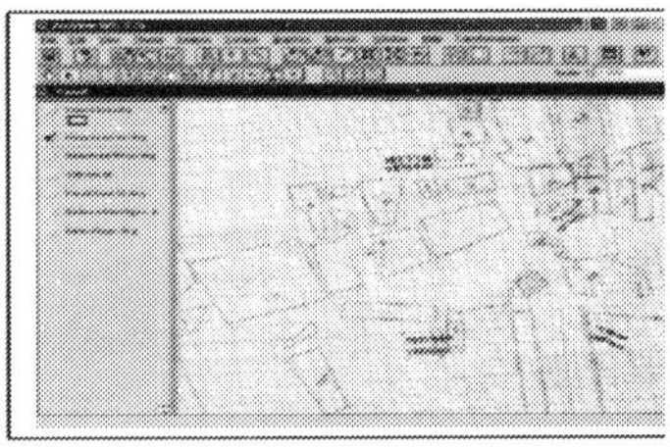

(GPS survey points were plotted in a graph paper and the layout was prepared) Geo Referenced survey blocks are in the in Arc view

Fig.8a Survayed Blocks overlaid on LISS IV image (Blue Colour)

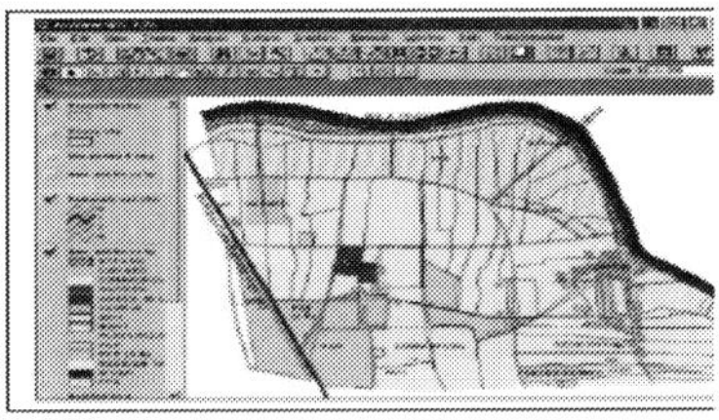

Fig.8b Survayed Blocks overlaid on Draft DP (Green Colour)

were under public and semi public land use class. The newly constructed apartments and apartments under construction were located in the residential land use class. The area, which designated as a Playground in the draft DP, was only vacant land during the survey. The Biotechnology Park in the proposal was under agricultural use at the time of survey. There was no construction in the green belt area close to the river (with in 60 m distance) except the Gaothan. When the digitized DP along with survey numbers overlaid on the image, it gives a detailed look of land conversion and other land use details, which can be added to the information system, for revenue collection and be a part of LIS of Balewadi.

The study proved that, the multi spectral LISS IV with 5.8 m resolution, it is possible to map the urban feature at 1:10,000 scale. With synoptic coverage, timely, accurate and cost effective data, LISS IV data is an ideal tool for generating land use information and can be efficiently used for monitoring and implementing the Development Plans sustainably. In this study, the method followed was very simple and easy, so that municipal corporation can adopt for their implementation process. Thus optimal management and better utilization of available resources is very much possible with the integrated approach.

Acknowledgements

The authors are grateful to ISG for having given an opportunity to write this article. Mr. R. Bhos and Mr. S. Paradesi are gratefully acknowledged for their full support in the conduct of field survey. People from Balewadi village deserve sincere thanks for their full co-operation. The authors are thankful to Mr. V.G. Dixit, NDA, for his encouragement and support.

REFERENCES

Census (2001), Census Department, Pune: Municipal Corporation.

Kalyanaraman.K., (2005). Large Scale Mapping using High Resolution Satellite image. ISG Newsletter, 11, (2 and 3).

Krishna Murthy, Y.V.N. and Adiga, S., (2000). Integrated Sustainable Development of Land and Water Resources using Space Technology Inputs. Space Forum, Special Issue on Indian Space Program.

Shekhar, S., (2004). Urban Sprawl Assessment: Entropy approach, GIS Development, 8 (5).

Shekhar. S., (2005). Monitoring the Growth of Pune by using Remote Sensing and GIS Techniques. Thesis Unpublished Thesis. Pune: University of Pune.

Vaisali Nandan, (2004). Municipal GIS: The Gorakhpur Experience.

http:/ /www.gisdevelopment net/ application /urban/ overview/ mi03224.htm

Patkar, V.D., D. Sampath Kumar, (2001). Mumbai Metropolitan Region Planning with the Aid of Geo-referencing Technologies. CARG-2001: Workshop cum Tutorials.

APPLICATIONS OF GEOSPATIAL TECHNOLOGY
A. Ganesh (Ed.), (2006) : pp. 195-203
Satish Serial Publishing House, New Delhi
E-mail: ssp@ndf.vsnl.net.in, hkjain1975@yahoo.com
website: www.satishserial.co

14

GIS Applications in Integrated Rural Development

N. D. MANI

Planning for integrated development of rural area is a stupendous task. It requires data/information about resources, problems and felt needs of the people. After the enhancement of 73rd constitutional amendment, the elected panchayat members also require data/ information for planning. The problem in the present scenario is information, which is often outdated, lacks spatial perspective, and takes long time for collection and collation and onward transmission. One of the new technologies in the age of information technology available to planners is geographical information system (GIS). It is a spatial information system having powerful set of tools for collecting, storing, retrieving, at will, transforming and displaying spatial data from real world for specific purposes (Burrough, 1986).

GIS has been used by variety of disciplines, which deal with spatial problems. GIS, therefore, is a convergence of technological and traditional disciplines. Among various disciplines geography, cartography, remote sensing, computer science etc are the important ones. Thus the users also contribute in improving the capabilities of GIS.

Literature Review on GIS Applications in India

A brief review of work completed in the area of GIS is given in the following paragraphs. Information is one of the basic requirements of planning and development. Therefore, a few attempts were made to furnish information for planning purposes at District/ Block levels.

Planning and Development

GIS was designed for a village in 1:2000 scale as a pilot project Sundaram et.al., (1996) and used for socio economic transformation and planning for service centres of a taluk in Madhya Pradesh (Vijay and Kaur, 1997). A map of Mallapuram District under district planning series was prepared using GIS (Sarkar and Nandi, 1998). A GIS was designed for Thanjavur and Salem District (Siddiq, et.al., 2000).

Watershed Development and Management

The tools of GIS were also applied in watershed planning and development. For watershed development, PAMAP GIS was used to draw maps (Jothi, et.al., 1996). ARCVIEW software was used to prepare watershed maps, classification of watershed into micro watersheds with attribute data (Ghouse and Kalyani, 1998).The overlay capability of GIS was used for preparing an action plan for land and water resource development at micro watershed level (Palaniswamy, et.al., 2000) and a priority map for a micro watershed were generated using GIS (Rusthum, et.al., 2000).

Urban Studies

The quality of urban environment of Kanpur town was assessed and the town was classified into good, moderate and poor by using IDRISI software (Sukumar et.al., 1999). Guwahati city was categorized into five suitability zones as high, medium, low, very low and unsuitable for conversation for futuristic and scientific planning (Sudhakar, 1999). The problem of flooding, chocking of drainage system of a part of Indore city were mapped using GIS (Ozha, et.al., 1998). The dispersion of sewerage out falls and simulation results after their release in the Arabian Sea were studied with the application of GIS (Vyas and Singh, 1999). The problem of water logging in bara track area was solved with GIS (Parekh and Patel, 2000). Urban sprawl of Chanakyapuri area in Delhi (Subankhan, et.al., 1998), growth of Coimbatore city in Tamil Nadu state (Sukumar, 2000) and Tiruchirappalli urban area (Samson and Annapurna, 2000) were analyzed with the tools of GIS. Hotspots of occurrence of crime were identified, spatio temporal patterns of crime and criminals were analyzed in Chennai city (Jaishankar and Shanmuga Priya, 2000).

Preparation of Maps

In India a group of people use GIS primarily for drawing maps while others use it as a versatile tool employing the entire gamut of its capabilities. Several studies were reported using the mapping capabilities of GIS. Using PC ARCInfo and PC-TIN software, various thematic maps have been overlaid and 3D modeling has been accomplished for analyzing reverine pollution monitoring (Ramalingam, et.al., 1998). CARIS-GIS software was used to digitize thematic maps (Mahamaya et.al., 1998), agriculture potentiality maps at village level for a mandal in Andhra Pradesh (Bhar and Sengupta, 1998), and land capability map covering four taluks in two districts of Gujarat State (Subramanian, 1998) were prepared using GIS as a mapping software. There were sporadic attempts made to modernize designing of cadastral maps (Dhal, et.al., 1998), designing enumeration blocks map for Census Operation (Das, et.al., 1998), and preparing cadastral maps (Bhagra, et.al., 1999) applying GIS techniques.

Environmental Studies

Environmental problem of water logging and salinity was analyzed using GIS in South West Punjab (Singh, 1998). GIS was employed for mapping of environmental quality and distribution of water borne diseases in Ahmedabad (Desai and Parjapati, 1998). IDSISI software used for mapping tannery polluted area in Dindigul Town (Mari, 1998). The ambient air quality was spatially analyzed using Arc- Info software in Pondicherry union Territory (Mouliswaran, et.al., 2000).

Natural Resource Management

Planning and management of resource is an important area for the application of GIS. ILWIS GIS has been employed to identify areas of potential resource at blocks level (Jacob, 1996). Mapping the changes in wastelands of Gurgaon District was conducted using GIS (Khan, et.al., 1999). A micro level action plan was prepared for the development and management of wastelands at Block level (Bhuraneswari, 1999). Separate management plans were formulated for land, water and human resources for Sangod Block in GIS environment (Rajashree, et.al., 1999). Attribute data models, topology overlays for cartographic visualization of marine pollution at global and national levels were prepared (Singhi and Vyas, 1999).

GIS was used for the assessment and management of mangrove forest at Pitchavaram wetlands (Gnanappazham, 1998) and wetlands in Bangalore City (Rajinikanth and Ramachandra, 2000). It was also employed for management of natural resources at District level (Perumal, 2000).

Ground Water Management

GIS was applied for mapping of groundwater for the assessment of potentiality and identification of areas of recharge. The quality of ground water in Madurai Metropolitan area was assessed (Rajmohan, 2000), water prospect map of Dhanbad watershed has been prepared (Srivatsava, 2000). Athoor Block in Dindigul District was divided into five major ground water potential categories such as very good, good, moderate, moderate to poor and nil (Motas and Ramalingam, 2000). Spatial variability in ground water was mapped for Erode District using GIS (Ismail, et.al., 2000). In another study, ground water potential zone maps were prepared for a part of Vellore Taluk in Vellore District (Porchian, et.al., 2000). The controlling parameters were identified in detecting the potential ground water targets (Palanivel and Ramasamy, 2000). The morphological parameters which are required for many hydrological studies have been evaluated using GIS (Jain and singh, 2000). It was employed for detecting natural recharge windows and causative factors of natural recharge (Ramasamy, 2000). Ground water resources were evaluated by adopting a multi criteria analysis using GIS (Balasubramaniam, 2000). GIS was used to describe the road environment under which bus accidents occur often on NH 47 (Ramasamy, 2000).

Development of GIS software

GIS software was developed for displaying DVD-1 file by Survey of India (Prasad, 1999).

The above review of literature reveals that GIS has a lot scope for use in solving the problems confronting the country. In the present study an effort has been made to design a GIS for a project 'Hunger Free Area Programme', implemented by the Faculty of Rural Development, Gandhigram Rural Institute.

Hunger Free Area Programme

Hunger Free Area programme was started as a pilot project in a village Panchayat - Kovilur Panchayat - in Vedasandur Panchayat Union of Dindigul district. It was funded by Department of Social Welfare, Government of Tamil Nadu and coordinated by M.S.Swaminathan Research Foundation, Chennai.

The programme intended to fill the areal gaps in the reach of the development programmes besides addressing the problems so far not covered by any of these development programmes. It was aimed at conducting a detailed study on various aspects related to hunger and preparing a profile of the study area i.e. Vedasandur block in Dindigul district.

As the formulation of action projects is a voluminous and laborious work, a GIS was designed to furnish information to the project implementer - Faculty of Rural Development.

Salient features of GIS

- It has been designed for the entire Vedasandur Panchayat Union, Dindigul District.
- IDRISI- a raster based software has been used.
- It is capable of supplying information at panchayat union and village panchayat level in 1: 50000 scale.
- There are 80 layers in the GIS. Of these layers, two layers are base maps, four layers show natural features, another four layers depict irrigation particulars, eight layers provide land use details, 25 layers furnish live stock information, 31 layers present human resources and the remaining six layers represent infrastructural facilities.
- In the designing of the GIS, Corel capture, Coral trace, Coral draw, ACAD 14, TOSCA 212 and IDRISI software were used.

Areas of Application

Currently, it is being used for identification and selection of villages of the programme. It provides basic information like, availability and accessibility of infrastructural facilities, concentration of SC/ ST population, work force etc.

Conclusion

The GIS is used for taking decisions relating to selection of villages for increasing the coverage of the project. There is a lot of scope for application the GIS in this programme as the project is extended to other village Panchayats of Vedasandur Block in Dindigul District.

REFERENCES

Balasubramaniam, A., (2000). GIS in Ground Water Development, Geoinformatics-2000, Coimbatore, pp 244 – 247.

Bhagra,K.C., et.al., (1999). Cadastral Survey Conventional to Modern Mapping: GPS, Photogrammetry, Photomaps and GIS Approach, Indian Cartographer, 19, pp. 68-73.

Bhar, S., and Mahuya Sengupta, (1998.) GIS for Maheswaram Mandal, Andhra Pradesh, Indian Cartographer, 18, pp. 342-344.

Bhuvaneswari, S., Rehman Ali and M.Asif, (1999). GIS and Remote Sensing for Integrated Management of Wastelands- A Case Study of Ferozepur Jhrika (Gurgaon), Hariyana, Indian Cartographer, Dehradun, 19, pp. 353-357.

Burrough, P.A., (1986). Principles of Geographical Information System for Land Resources Assessment. New York: Oxford University Press.

Das, D.N., and Krishna Kumar, (1998). Enumeration Blocks in Census Operations: Scope for GIS Application, Indian Cartographer, 18, pp. 324-328.

Desai, A., and Avantika Prajapati, (1998). Mapping of Environment Quality and Distribution of Water Borne Diseases with GIS: Case Study of Ahmedabad, Indian Cartographer, 18, pp. 212-217.

Dhal, N., S. Ravi aNd J.K. Rath, (1998). Total Technology Solution for Modernization of Cadastral System Leading to Parcel Based LIS, Indian Cartographer, 18, pp. 320-323.

Falguni P. Parekh and H.M.Patel, Geographical Information System Applications in Irrigation Management, Geoinformatics-2000, Coimbatore, 2000, pp.233-237.

Ghouse, M. S., and Kalyani, A., (1988). Application of ArcView 3A GIS Software to Study Srivilliputhur Watershed in Tamil Nadu. Indian cartographer 18, pp. 177-180.

Gnanappazham, L., (1998). Application Remote Sensing and GIS in Management of Coastal Area, Indian Cartographer, 18, pp. 320-323.

Hari Om Prasad, (1999). Graphical display of DVD-I Data and PC Platform-An Indigenous Approach Toward GIS, Indian Cartographer, 19, pp. 74-77.

Ismail, M.K., S, K.A. Hemalatha, M. Prasanthidevi, M. Radhika, S. Pattabi and S. Balasubramanian, (2000). Mapping of Spatial Variabilty of Ground Water in Erode District using GIS, Geoinformatics – 2000, Coimbatore, pp. 158-165.

Jacob, D. (1996). Application of GIS for Natural Resource Data Management System of Thiruvananthapuram District Indian Cartographer 16, pp. 231-233.

Jain, S.K., and Singh R.D., (2000). Estimation of Morphological Parameters for small Catchments Using GIS, Geoinformatics -2000, Coimbatore, pp. 136 -143.

Jaishankar, K., and Shanmugapriya.S., (2000). Application of GIS in Indian Policing: A Study of Chennai, Geoinformatics-2000, Coimbatore, pp. 90-115.

Joshi, S.C., O.P. Gaba and Sunil Chandra, (1996). Forest Resource Management Indian Cartographer, 16, pp. 211-220.

Khan, S., S. Rehan Ali and Mohd Asit, (1998). Application of GIS in Urban Sprawl Mapping and Monitoring- A Case Study of Chanakya Place, Delhi, Indian Cartographer, 18, pp. 218-224.

Khan. S., S.R. Ali, M. Asif and W.A. Zaidi, (1999) Wasteland Change Analysis of Gurgaon District in Haryanan using Remote Sensing and GIS, Indian Cartographer, 17, pp. 290-297.

Mahamaya, C., K. Soman, P.P., Quesph and C. Sakunthala, "Application of GIS in Riverine Pollution Monitoring - A Case Study on Killiar Basin South Kerala," Indian Cartographer, 16, pp. 199 - 202.

Mani,N.D., (1998). Mapping of Tannery Polluted Area, Indian Cartographer, 18, pp. 245-249.

Moses E.J., and Ramalingam, M; (2000). Groundwater Development in Western Ghats Region Using Remote Sensing and GIS, Geoinformatics-2000, Coimbatore pp. 144 -148.

Moulishwaran. C.R, et.al., (2000). A GIS study on Spatial Distribution of Ambient Air Quality in Pondicherry. Geoinformatics – 2000, Coimbatore, pp. 158-165.

Ozha, M.K., Devender Kaur and T.A. Sihoriwala, (1998). Application of GIS and Space Borne Data in Management and Planning of Drainage Problem- A Case Study, Indian Cartographer, 18, pp. 193-194.

Palanisamy, K., Paramasivam and Lakshiminarayan, (2000). Geographical Information System based Decision support for Annur sub watershed planning" Geoinformatic 2000, Coimbatore, pp. 48-59.

Palanivel,K and Ramasamy, SM., (2000). GIS and Hard Rock Acquifer Function Modelling in Western Ghats Region ,Tamil Nadu, India, Geoinformatics-2000, Coimbatore, pp. 205 – 210.

Perumal. A, (2000). Remote Sensing and GIS: Application Technologies for Natural Resources Management in Natural Mission, Geoinformatics-2000, Coimbatore, pp. 1-5.

Porchlan, P. Ramkumar, T and V. Lazer John. (2000). Identification of Ground Water Potential Zones using Remote Sensing and GIS – Case Study from part of Vellur Taluk Geoinformatics 2000, Coimbatore, pp. 185-190.

Rajashree, V., Bothale and I.R. Sharma, (1999). Sustainable Resource Management in Sangod Block of Kota Tehil using Remote Sensing and GIS, Indian Cartographer, Dehradun, 19, pp. 301-304.

Rajinikanth, R., and Ramachandra, T.V., (2000). Effective Wetland Management Using GIS, Geoinformatics 2000,Coimbatore pp. 262 – 275.

Rajmohan, S., (2000). GIS Application in Groundwater Quality Assessment in Madurai Corporation, Geoinformatics–2000,Coimbatore, pp. 127-135.

Ramalingam, M., et.al., (1996) Brachish Water Aquaculture Site Selection using Remote Sensing and GIS under Coastal Zone Management" Indian Cartographer 16, pp.164-168.

Ramasamy, M., et. al., (2000). Nourishment of Acquifer Systems through Natural Recharge Using GIS, Geoinformatics -2000, Coimbatore, pp.211 – 217.

Ramasamy, S.M, (2000). Geoinformatics and the Visualisation of Environmental Systems of Southern part of the Indian Peninsula, Geoinformatics-2000, Coimbatore, pp. 197-204.

Rusthum, S., Kesava Rao, P. and Muralikrishna, (2000). Remote Sensing Application and GIS for Watershed Prioritisation and Management, Geoinformatic 2000, Coimbatore, 2000, pp.60-64.

Saddiq, Z., Jarshankar, J., and Ravindran, A.,(2000). GIS for Governmental Administration, Geoinformatic 2000, Coimbatore, pp.228.

Samson, M., and Annapurna, (200).Integrated GIS as a Tool for Urban Land Use Monitoring and Planning, Geoinformatics 2000, Coimbatore, pp. 71-78.

Sarkar, A.K., and S. Nandi, (1998). DPMS Malappuram District an Experiment in Digital Mapping, Indian Cartographer. 18, pp. 392-394.

Singh, K.P., (1997). Investigation of Environmental Problem of Water Logging and Salinity -Cartographic and GIS Approach, Indian Cartographer, 17, pp. 219-224.

Singhi P., and Lakshmi Vyas, (1999). Spatial Analysis and Mapping of Marine Pollution – A Global Perspective with Specific Reference to the Emerging Hot spots of Indian Sea Water, Indian Cartographer, 19, pp. 325-332.

Srivastava, V.K., (2000). Water Resources Management through Remote Sensing and GIS: A Case Study of Dhanbad Watershed, Geoinformatic 2000, Coimbatore, pp. 238-243.

Subramanian, c., (1998). Mapping of Land Capability Using GIS, Indian Cartographer, 18, pp. 371-375.

Sudhakar. S., Shantanu, K. Dutta and Amit K. Bhattacharya, (1999). Updation of Infrastructure Facilities and Land Suitability Analysis of Guwahati City and its Environs- RS and GIS, Indian Cartographer, 19, pp. 220-222.

Sukumar, A., B. Sukumar and B. Arunachalam, (1999). Urban Environmental Quality Assessment for Kannur Town using Remote Sensing Data in GIS Environment, Indian Cartographer, 19, pp. 143-145.

Sukumar,B. (2000). Application of Geographic information System for Transportation Planning and Management, Geoinformatics-2000, Coimbatore, pp.79-85.

Sundaram, A., M.K. Palanisamy and V. Guruswamy, (1996). "Land Information Management for Sevilimedu Panchayat Rural area in Tamil Nadu through Geographical Information System", Indian Cartographer, 16, pp. 137 - 143.

Vijay Phase and Devender Kaur, (1997). Application of GIS in Socio Economic Transformation and Planning of Service Centres of Tribal Kukshi Tahsil of Dhar District, Indian Cartographer, 17.

Vyas L., and Premlata Singh, (1999). Application of GIS for Monitoring the Dispersion of Sewage out falls along the West Coast of Mumbai, Indian Cartographer, 19, pp. 356-361.

APPLICATIONS OF GEOSPATIAL TECHNOLOGY
A. Ganesh (Ed.), (2006) : pp. 205-217
Satish Serial Publishing House, New Delhi
E-mail: ssp@ndf.vsnl.net.in, hkjain1975@yahoo.com
website: www.satishserial.co

15

GIS for Micro Level Planning

N.D. MANI

The Governments both at the Central and the State level formulate and implement development programmes in rural areas. These programmes fail to make expected impact on the rural areas as these programmes fail to consider the local resources and requirements. One of the reasons attributed to this sorry state of affairs is non-availability of information about rural areas.

Any micro-level planning exercise requires systematic and scientific collection, documentation and representation of relevant data for present and future use. There is a need for a comprehensive information system for storage, retrieval and processing for planning at the block level. In this context, an attempt has been made to study the available information systems for block level planning, review available data at block level and collect opinion of Panchayat Presidents and design an information system for a block suited to the requirements of the potential users viz, Panchayat leaders at village Panchayat/Panchayat union leaders, academicians, researchers etc.

Study on Information Systems

The information systems established by the government, voluntary Agencies and the academic institutions shows that the Rural Knowledge Centers of MSSRF and Telecentre of TANUVAS are being manned and used by the local villagers as these centers are located in their villages and also they supply the data/information required by them. On the other hand PRIS of IRS and Melur project of Government of Tamil Nadu and Priasoft, Ruralsoft and Ruralbazar are located at the Collectorates. The software is used primarily by the district bureaucracy and to a lesser extent by the villagers because these facilities are not easily accessible. Thus, it is clear that there is a need for a comprehensive information system located at villages for providing the data collected by the Government Departments, satellite images, maps of Survey of India etc., as required by the villagers.

Opinion of Village Panchayat Presidents

For the purpose of collection of primary data an interview schedule has been prepared. This field-tested interview schedule was used for collecting data from 22 village Panchayats in a face-to-face paradigm. It was prepared with a view to collecting data related to general information about the Panchayat Presidents, role of information in Panchayat planning, data required for decision making, use of natural resources, sources of data, availability of adequate and latest data etc. Besides, awareness about computer center of National Information Centre (NIC) and software of NIC, problems faced by them during data collection and their suggestions for solving them were also collected.

The major findings of the analysis of data collected from 22 panchayat Presidents are:

General Information

- More than 45 per cent of the Panchayat Presidents are aged between 41 and 50 and another 36 per cent between 31 and 40 years. Half of them educated up to10th standard.
- The family size of nearly 82 per cent of the Panchayat Presidents is less than five members.
- As far as occupation is concerned more than three- fourth of them are farmers and less than one-fifth of them are housewives.

Use of Data in Decision Making

- The Panchayat Presidents are making decisions on daily basis based on the urgency in areas like drinking water, road, repairing of hand pump, drinking water pipe line and drainage facilities.
- On long term basis, they take decisions related to road, drainage facilities, school, public toilet etc.,
- The Panchayat Presidents take decisions in these fields based on the size of population demanding these facilities (95.5 per cent), and the data available at local schools (18.2 per cent).

- The major areas of decision making include facilities like drinking water, drainage, public toilet etc. as their planning areas is a hamlet they were able to manage with the data available in their office, furnished by the local people and school teachers.

Use of Natural Resources

- All the respondents said they make use of natural resources available in their Village Panchayat.
- Tamarind trees and tanks were auctioned for generating income for the panchayat, under groundwater and soils were tested for agricultural purposes.
- They identified these resources with the help of Government officials, water diviner and villagers.

Sources of Data

- The Panchayat Presidents identified the resources by using the data collected from Collectorate (45.6 per cent), District Rural Development Agency (86.4 per cent), Panchayat Union Office (100 per cent), Village Panchayat (72.7 per cent) and local people (32 per cent).
- According to 13.6 per cent of Panchayat Presidents, the data required by them are not available from the above sources.
- All these 13.6 per cent Panchayat Presidents manage the situation, by manipulating the available data, passing new resolutions in the Panchayat meeting according to the availability of funds and depending upon local conditions.
- With regard to sources of data, 82 per cent of them opined that the required data are not available from single source and therefore they have to depend on other sources for data.

Adequacy of Data

- In the opinion of 36 per cent of Panchayat Presidents, the data collected by them are inadequate. They manage the situation, by manipulating the data collected by them, passing new resolution in place of old ones and manage it in consultation with the members.
- Of the total respondents, 91 per cent of them could not collect the latest data. They managed with old data and by meeting concerned officials in the concerned offices.

- Eighty six percent of the Panchayat Presidents could not collect the required data on time. The Presidents used to postpone the decision-making till they collect the required data (68 per cent), while some of them pass resolutions according to available data.

NIC Computer Centre

- None of them were aware of functioning of computer centre managed by NIC.
- They are also not aware of softwares like Ruralbazar, Ruralsoft 2000, Priasoft and Touch Screen Information Kiosk.

Problem faced during data collection

- The problems faced by the Panchayat Presidents during data collection include non-availability of data (91 per cent), repeated visits for collecting existing data (77 percent) and availability of outdated data (73 per cent).

Suggestions

- Separate information centre is required (77 percent)
- It should be manned by trained officials (73 percent)
- Latest data should be collected and stored in the information centre (68 per cent)
- Once in a year Panchayat-wise data should be collected and stored in the information centre.

Information System

- All the respondents are for starting a new and separate information system for Athoor Block i.e. within the block.

Participants in the Information System

- Panchayat Presidents will support and cooperate with Gandhigram Rural Institute (77 per cent)
- Panchayat Presidents will furnish data (64 per cent)
- Panchayat Presidents will use the data for decision making (50 per cent)

Collection of Secondary Data

A study was undertaken about the rural development programmes implemented in Athoor Panchayat Union of Dindigul

District in order to prepare a list of data required for the better decision making at block and village Panchayat level. Based on this, the data required for the better implementation of these rural development programmes were collected from the following Offices.

Sources of Data

S.No.	Source of Data	Theme of Data	Level of Data	Year of Data
1.	Census of India, New Delhi	Population	Revenue Village	1991
2.	Asst. Director of Statistics	Land use	Revenue Village	2001
3.	Asst. Director of Statistics	Irrigation	Revenue Village	2001
4.	Asst. Director of Statistics	Pump sets and Tractors	Revenue Village	2001
5.	Dept. of Animal Husbandry	Animal Husbandry	Revenue Village	2001
6.	Panchayat Union Office, Athoor	Infrastructural facilities	Panchayat	2000
7.	Panchayat Union Office, Athoor	IRDP	Panchayat	1994-97
8.	Panchayat Union Office, Athoor	Jawahar Rozgar Yojana	Panchayat	1991-96
9.	Survey of India Maps	Contour Slope Drainage Railway Road Network	1:50,000	
10.	Survey and Land Records Dept.	Revenue Village Boundary	1:50,000	

Databases Created in Access

Following databases were created from the data collected from the above sources for their use in the proposed Information System for Panchayat Planning for Block Level.

Databases Created

Theme of Data	Name of the File
Population	Popu 1991
Land Use	Landuse
Irrigation	Irri2001.
Pump sets & Tractors	Pump2001
Cow	Cow2001
Buffaloes	Buff2001

Sheep	Sheep2001
IRDP	IRDP9596
IRDP	irdp9495
Jawahar Rozgar Yojana	JRY9196
Infrastructural facilities	Infra 2000

Awareness Programme Conducted

An awareness programme about the use of information in Panchayat Planning was organized at Gandhigram Rural Institute on 15.5.2002. The Village Panchayat Presidents were participants of the awareness programme. The aim of the programme was to explain the process of decision-making and the role of reliable and recent data in decision-making. In addition, a demonstration on how the computer technology could be used in supplying information, locating records in a database and sending of information using e-mail to the higher authorities. They were also exposed to Geographic Information System Technology and all of them were given Athoor Block Map.

For the purpose of creating awareness, lectures were delivered and printouts of the lectures were given to them in Tamil. During the discussion session, Panchayat Presidents came forward to supply the data for entering and storing in the database of the proposed information system.

The major gaps listed during the discussion by the Panchayat Presidents are,

- They were unable to get the required data-for e.g. SC population
- The officials withheld the information related to the developmental/welfare programmes
- The Panchayat Secretaries are not sharing even the essential particulars for decision-making. For e.g. Minutes of Panchayat Meetings.
- In their opinion, the prerequisite for the establishment and use of computerized information system for Panchayat Planning is creating awareness and providing training to Panchayat Secretaries.

In addition to reporting formats, the documents and registers maintained as mentioned below in the office of the Village Panchayat, were studied in detail. The format for entering the data generated at village panchayats' level has been prepared in MS-Access. These MS

Access databases can be accessed in VB based software prepared exclusively for this purpose.

- House Tax Demand Notice.
- House Tax Receipt Book
- Pending house Tax demand Register
- Profession Tax Receipt Book
- Profession Tax demand Register
- Miscellaneous Receipt Book
- Panchayat Tax and other tax collection Register
- Subsidy Grant of Panchayat and others earmarked income Register
- Chitta
- Pending Collection Demand Register
- Cash Book
- Panchayat Advance Register
- Approval Statement Register-Head-Wise
- Fund Receipt Register
- Estimate and Allocation of Fund for the work
- Assets and Tools Register
- Maintenance Register
- Stock Register
- Voucher
- Cash Receipt
- Stock Register of unserviceable article
- Project Work Stock Register
- Cash Book
- Government Subsidy classification Register
- Estimate and Allocation of fund for the work carried out by the Panchayat
- Cash Book of Projects
- Income and Expenditure Register of Project Accounts-head wise
- Project Work Register
- Project Beneficiaries Register
- Monthly Accounts Form
- Financial Statement of Account
- Resolution Book
- Resolution Copy
- Resolution Copy (Additional)

Creation of Information System

The databases created based on data collected from Government Departments, NIC, Panchayat Union Office, Maps of Survey of India etc. and MS Access databases formed the base of Information System for Panchayat Level Planning. Thus the information system was evolving as per the requirements of users.

For continuous flow of data from panchayats, the Secretaries of all Village Panchayats have been trained in data entry in the formats created in MS Access based on the analysis of documents and registers maintained by them.

Maps Prepared

Though a set of 50 maps prepared only following maps have been taken as hard copies from GeoMedia Pro-GIS-Software.

1.Slope (Fig. 1)
2.Soils (Fig. 2)
3.Drainage (Fig. 3)
4.Quality of Ground Water (Fig. 4)
5.Depth of Ground Water Table (Fig. 5)
6.Flow Irrigation (Fig. 6)
7.Landuse / Land cover (Fig. 7)
8.Net Sown Area (Fig. 8)

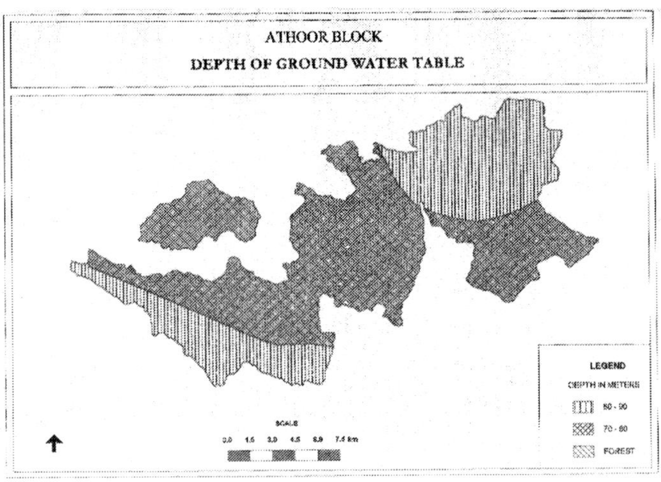

Fig.1

9. Current Fallows (Fig. 9)
10. Density of Population (Fig. 10)
11. Sex Ratio (Fig. 11)
12. SC Population (Fig. 12)
13. Roads and Railways (Fig. 13)

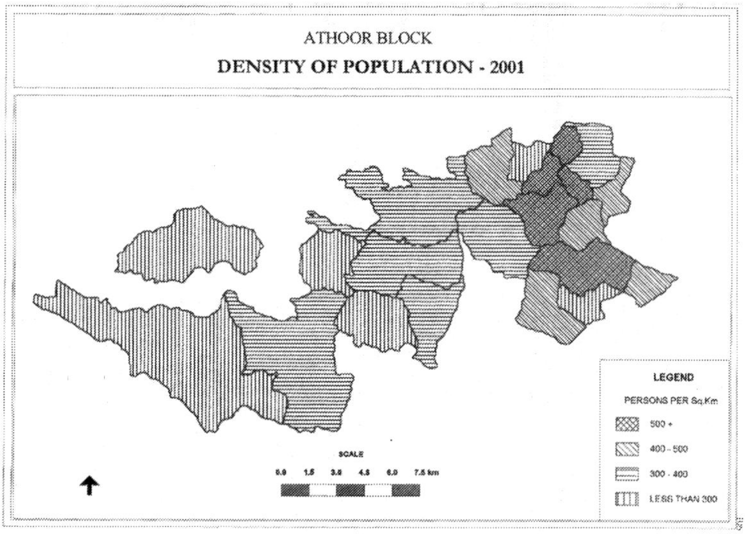

Fig.2

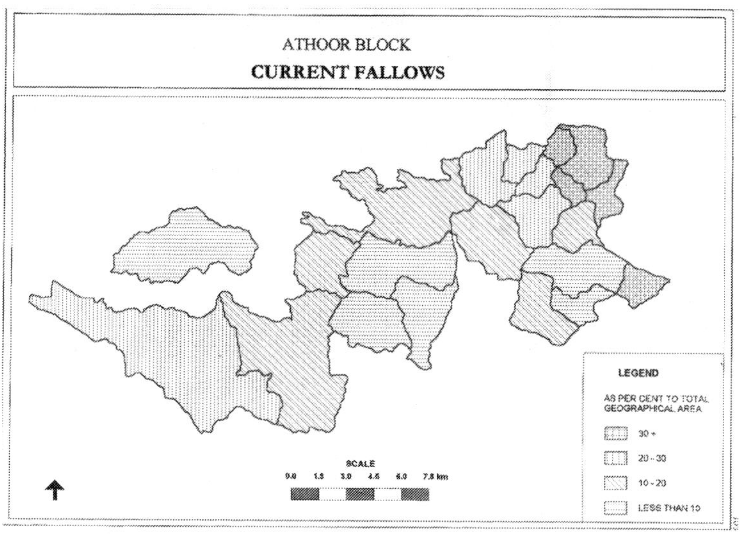

Fig.3

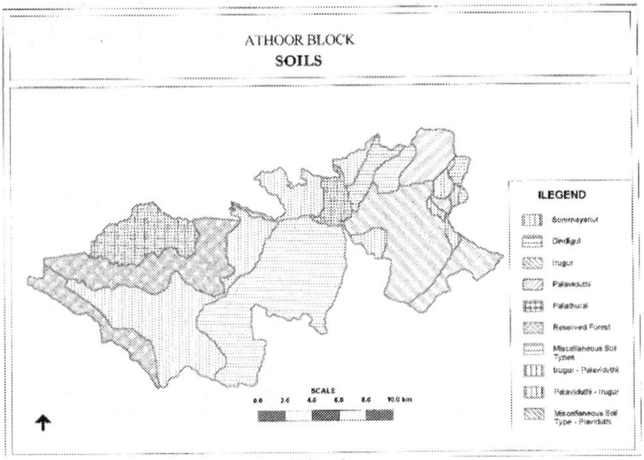

Fig.4

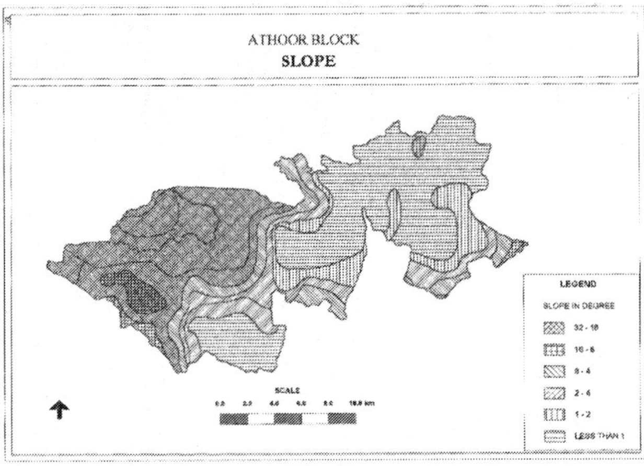

Fig.5

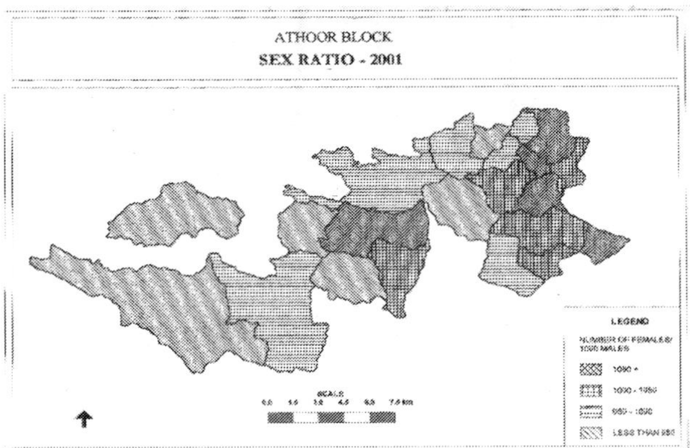

Fig.6

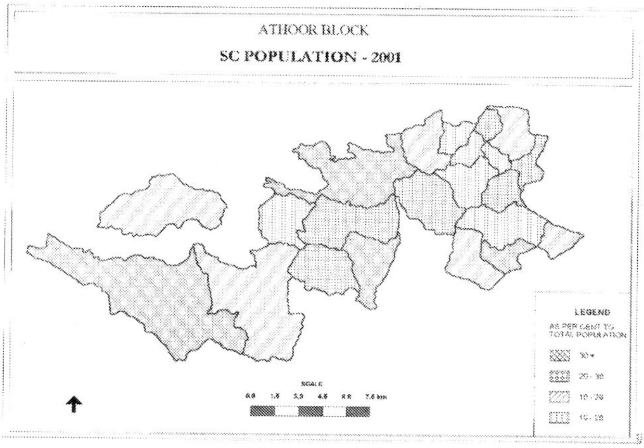

Fig.7

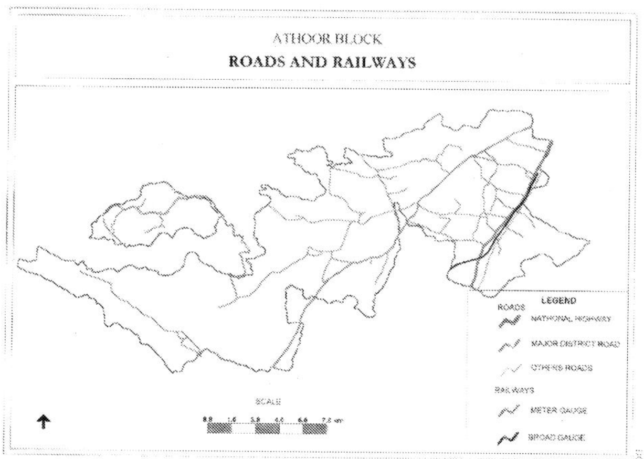

Fig.8

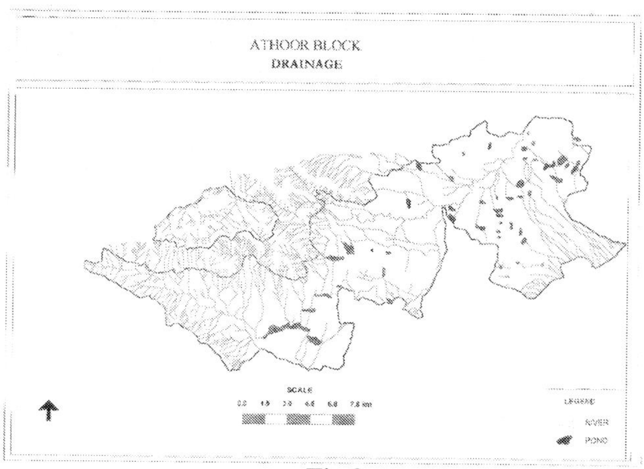

Fig.9

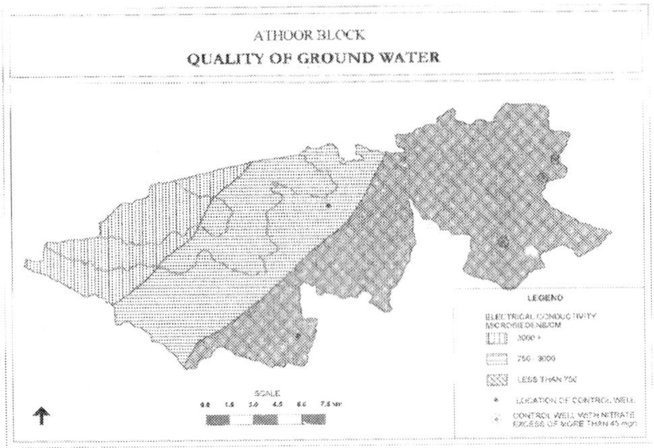

Fig.10

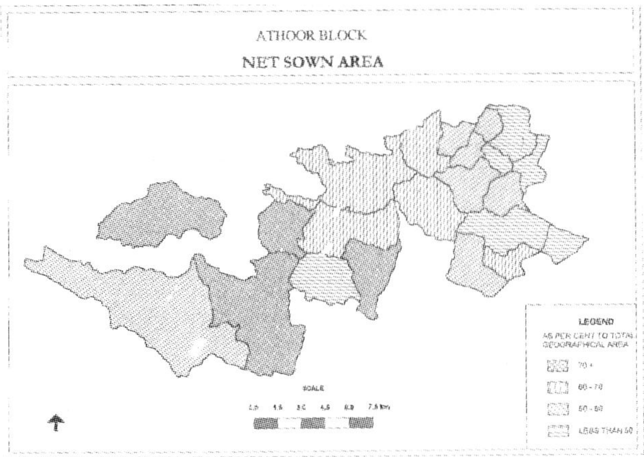

Fig.11

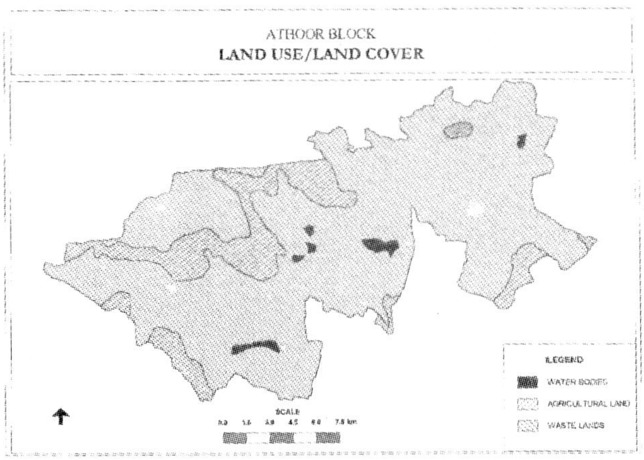

Fig.12

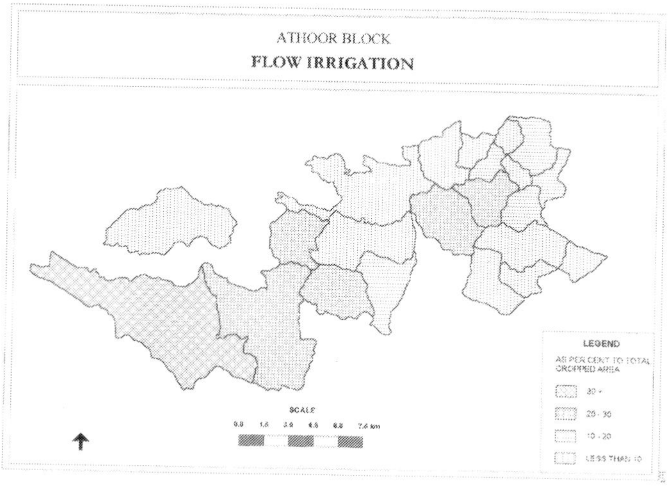

Fig.13

At present, the Geographic Information System of Athoor Block is being used by the village panchayats, block bureaucracy, local non-governmental agencies and research scholars.

REFERENCE

Mani, N.D., (2003). Designing an Information System for Planning at Block Level, New Delhi: UGC Project Report.

APPLICATIONS OF GEOSPATIAL TECHNOLOGY
A. Ganesh (Ed.), (2006) : pp. 219-225
Satish Serial Publishing House, New Delhi
E-mail: ssp@ndf.vsnl.net.in, hkjain1975@yahoo.com
website: www.satishserial.co

16

Crime Modeling and GIS

K. BANUKUMAR, P. RAMU, AND A. GANESH

The traditional and age-old system of intelligence and criminal record maintenance has failed to live up to the requirements of the existing crime scenario. Manual processes neither provide accurate, reliable and comprehensive data round the clock nor does it help in trend prediction and decision support. It also results in lower productivity and ineffective utilization of manpower. The solution to this ever-increasing problem lies in the effective use of Information Technology.

Geographic Information System (GIS) uses geography and computer-generated maps as an interface for integrating and accessing massive amounts of location-based information. GIS allows police personnel to plan effectively for emergency response, determine mitigation priorities, analyze historical events, and predict future events. GIS can also be used to get critical information to emergency responders upon dispatch or while en route to an incident to assist in tactical planning and response. GIS helps identify potential suspects to increase investigators suspect base when no leads are evident. GIS plays an important role in crime mapping and analysis. Response capabilities often rely on a variety of data from multiple agencies and sources.

The ability to access and process information quickly while displaying it in a spatial and visual medium allows agencies to allocate resources quickly and more effectively. In the 'mission-critical' nature of law enforcement, information about the location of a crime, incident, suspect, or victim is often crucial to determine the manner and size of the response. GIS software helps co-ordinate vast amounts of location-based data from multiple sources. It enables the user to layer the data and view the data most critical to the particular issue or mission. It is used world over by police departments, both large and small, to provide mapping solutions for crime analysis, criminal tracking, traffic safety, community policing, Intranet/Internet mapping, and numerous other tasks. GIS helps crime officers determine potential crime sites by examining complex seemingly unrelated criteria and displaying them all in a graphical, layered, spatial interface or map.

It also helps them map inmate populations, fixtures, and equipment to provide for the safety of inmates by separating gang members, identifying high-risk or potentially violent inmates, and identifying hazardous locations in an area. It reduces the potential for internal violence by providing better command and control. GIS functions, when combined with capabilities of location identification devices such as GPS facilitate tracking the movement of high-risk inmates or at-risk personnel throughout an area. It is more cost-effective for the crime analyst to come up with the information than for patrol officers to do it themselves. Virtually every operational activity in the police department includes spatial relationships. Traditionally, these activities have been supported by paper maps and pins.

Police officers now have the ability to immediately generate maps directly relevant to the situation at hand. Police agencies collect vast amounts of data from many sources including called-for-services, arrests, first information reports and daily report Data in this form, however, can be difficult to visualize. The same information displayed graphically provides a powerful decision making tool for investigators, supervisors, and administrators. The visual format shows relationships and patterns that are buried in the data. GIS could also be used to explore the relationship between crime and the environment.

Role of Internet/Intranet

Internet/Intranet capability enables an agency to serve various information to substations and remote offices, and to serve maps regionally to other agencies. It provides the ability to serve multiple applications from a single server to any number of clients, and serve dynamic maps and data quickly on a variety of servers. It also centralizes the administration of both data and applications, and can easily accommodate expansion of the system as the number of clients rises. Officers will be able to access several reports including summaries of activity for a user-specified period, lists of incidents occurring at a single address, and charts analyzing day-of-week and time-of-day relationships for aggravated assaults and robberies. Another benefit of the Intranet-based application will be the team's ability to turn around late-breaking information immediately.

It helps distribute the mapping functions to each of the district offices, enabling any officer or detective to generate customized maps. The more information officers have, the better equipped they are to do their job. The Intranet based application should be very secure and safe against unauthorized usage. To maintain the confidentiality of information, access to the system could be controlled through passwords.

Crime Mapping

Maps offer crime analysts graphic representations of crime-related issues. An understanding of where and why crimes occur can improve attempts to fight crime. Mapping crime can help police protect citizens more effectively. Simple maps that display the locations where crimes or concentrations of crimes have occurred can be used to help direct patrols to places they are most needed. Policy makers in police departments might use more complex maps to observe trends in criminal activity, and maps may prove invaluable in solving criminal cases.

Display Spatial Patterns of Events

Digital maps are the quickest means of visualizing the entire crime scenario. The locations of crime events, arrests, etc. can be routinely displayed on maps. This provides an easy method of viewing activities in an area rather than searching through a listing of events. Maps can also be used to convey more than one type of information at a time. Crime locations can be symbolized according to the day of week, type of crime, modus operandi (a particular suspect's method of operation when committing a crime) or frequency.

Integrate Community Characteristics

Community characteristics (for e.g., slums, markets, colleges, parks, alcohol permit locations, red light area, etc.) can be routinely displayed on maps while analyzing crime patterns to interpret relationship between these characteristics and the crime. For example, the locations of aggravated assaults, robberies and alcohol permits can be displayed to see if crime is clustering around locations that sell alcohol. Other mapping data such as bus routes and public housing can also be displayed at the same time to analyze relationships between neighbourhood characteristics and crime.

Produce Thematic Maps

Maps can be produced at any geographic level (e.g. Police stations, divisions, or zones) to aid in the analysis of crime patterns. Each response area can be shaded to represent the number of crimes that occurred in that area during a specific time frame. The darker the shade, the more events that occurred within the response area. These thematic maps can also be used to show the change in an area's crime rate. The percent change in the number of crime incidents can be displayed by shading each area according to whether there was an increase, decrease or no change.

Crime Analysis

Crime Analysis is defined as a set of systematic, analytical processes directed at providing timely and pertinent information relative to crime patterns and trend correlations to assist the operational and administrative personnel in planning the deployment of resources for the prevention and suppression of criminal activities, aiding the investigative process, and increasing apprehensions and the clearance of cases. It supports a number of department functions including patrol deployment, special operations, and tactical units, investigations, planning and research, crime prevention, and administrative services. Crime analysis can be divided into three categories:

Tactical: An analytical process that provides information used to assist operations personnel (patrol and investigative officers) in identifying specific and immediate crime trends, patterns, series, sprees and hotspots, providing investigative leads and clearing cases. Analysis includes associating criminal activity by method of the crime, time, date, location, suspect, vehicle, and other types of information.

Strategic: Concerned with long-range problems and projections of long-term increases or decreases in crime (crime trends). Strategic analysis also includes the preparation of crime statistical summaries, resource acquisition, and allocation studies.

Administrative: Focuses on provision of economic, geographic, or social information to administration.

Perform Radial Analysis: GIS can be used to measure the extent and type of problem within a certain distance around a particular location (e.g., street intersection, slum, bar, etc.). The distance can be anything that one wants to use as a radius around a particular location. For instance, we can quickly examine the type and number of offences and arrests that occurred within 500m of a particular location. GIS allows an analyst to quickly eliminate excess information. In addition to focusing on a single location, GIS can also be used to examine events that occurred within a certain distance of multiple locations. For example, an analyst can determine the number and types of offences that have occurred within 500m of all schools. A radial analysis can also be performed to determine which residents might need to be notified when a particular offender has moved into the area.

Identify Clusters of Events (hotspots): GIS identifies areas that contain dense clusters of events (hotspots). These high concentration areas usually demand special police attention. For example, GIS allows an analyst to identify all of the areas in a police station area where at least 5 robberies occurred within a 1km radius. These areas are then outlined on the map. Using GIS to identify hotspots provides a consistent method to measure concentrations of criminal events over time. Hotspots of violent crime, robbery, residential burglary, commercial burglary, auto theft, rape, etc. can be calculated every month for each police station area.

Compare Locations of Hotspots across Time: Crime hotspots that have been identified over several months can be displayed at the same time. This allows for the identification of areas with chronic problems and indicates the direction in which a particular crime may be shifting. These types of maps can also be used to solicit resources for an area from other public and/or private agencies.

Compare Hotspots of Different Crime Types: Hotspots of different offence types can be displayed to identify where they overlap. For instance, residential burglary hotspots can be displayed along with robbery hotspots to discover where they overlap. A more detailed analysis of these intersecting areas can then be performed.

Shade Grid Cell Maps: While multiple crimes (or events) at the same location are not visible on a pin map, they are clearly accounted for in a grid cell map. For instance, ten thefts occurring at a single location would be represented by a single point on a pin map. Using grid cell mapping, all ten thefts would be counted toward the total for the grid cell that would determine the shade of the cell. The darker the shade, the higher the number of incidents occurred at that particular location.

Analyze Multiple Variables: The relationship between one variable with another can be analyzed and this would indicate the strength and direction of the relationship between the two variables. For instance one could analyze the relationship between the number of alcohol permit locations and the number of fighting. Regression analysis is used to predict the value of a dependent variable such as violent crime rate based on the values of independent variables such as the poverty level within an area. Regressing allows for additional variables like illiteracy, backwardness, habits, etc. to be included in the model. The results of this type of analysis might be used to solicit resources from other public/private agencies to help reduce the crime rate.

Crime Investigation

Historically, the causes and origins of crime have been the subject of investigation. Some factors known to affect the volume and type of crime occurring from place to place are:

- Population density and degree of urbanization.
- Racial heterogeneity.
- Variations in composition of the population, particularly youth concentration.
- Stability of population with respect to residents' mobility, commuting patterns, and
- Modes of transportation and highway system.
- Economic conditions, including median income, poverty level, and job availability.
- Cultural factors and educational, recreational, and religious characteristics.
- Family conditions with respect to divorce and family cohesiveness.
- Climate.
- Effective strength of law enforcement agencies.
- Administrative and investigative emphases of law enforcement.
- Policies of other components of the criminal justice system (e.g. prosecutorial, judicial, correctional, and probational).

- Citizens' attitudes toward crime.
- Crime reporting practices of the citizenry.

GIS helps identify potential suspects to increase investigators suspect base when no leads are evident.

Investigate Serial Offences: The home addresses, work addresses, and places frequented by victims and suspects of serial offences can be displayed. When details of both suspects and victims are viewed in this manner, intersections between the two can be seen. The spatial patterns of less serious offences that often lead up to more serious crimes can be examined. This is especially important since serial rapists often begin their criminal careers committing other offences such as peeping tom incidents, indecent exposures, or residential burglaries.

Predict Behaviour: Based on the crime history, models can be developed to predict behaviour of criminals. Using these models, the probable location of an offender's home can be identified. The geographic area where the next crime is most likely to occur can also be predicted.

Identify Suspects: Crime data consisting of spatial patterns, suspect information and method of operation characteristics can be combined with arrest locations to provide investigators a tool to identify potential suspects. Investigators can quickly see hotspot locations and the people who have been previously arrested for the particular crime around these locations.

REFERENCES

Community Problem-oriented Policing using GIS (www.charmeck.nc.us)

Law enforcement and criminal justice (www.esri.com)

Crime Analysis (www.state.ia.us)

1995 Comparison of Iowa, Regional and National Crime Rates (www.ci.gilbert.az.us)

APPLICATIONS OF GEOSPATIAL TECHNOLOGY
A. Ganesh (Ed.), (2006) : pp. 227-231
Satish Serial Publishing House, New Delhi
E-mail: ssp@ndf.vsnl.net.in, hkjain1975@yahoo.com
website: www.satishserial.co

17

Role of GIS in Combating Crimes

O. M. MURALI

The current trends in crime related activities are increasingly becoming more complicated and mysterious in analysing and understanding due to rapidly changing lifestyle, socio-economic pressure, unemployment, migration, encroachment of urban set up onto rural areas, eroding human relationships, personal enmity, and competition. Everyday, hundreds of people are falling prey to criminals in cities. Some of them are reported and others remain unknown to the outside world.

In recent times, the incidence of rape, particularly in mega cities has become a regular phenomenon. In most of these cases, victims knew the criminals and in certain cases, victims are targeted in isolation in favourable geographic locations which are isolated and where there is no outside help possible. This type of crime is similar to that of highway robbery where victims may never get assistance when it is needed. Here, criminals flourish well owing to their time of attack which takes place during silent hours. From this, it indicates that there is a geographic preference for every criminal, who with his knowledge of local geography acts safely, remains undetected for long and escapes the detection and apprehension by police officials. Communal riots, highway robbery, kidnapping, murder, rape, theft and train loot are some of the crime types most of which happen in specific geographic locations. These are enhanced with the deep understanding of these locations by the criminals. In this way, criminals remain successful in their attempt to target the general public.

Social Causes of Crime

Incidents of crime happen in some specific geographic locations in cities and rural areas. These crimes are committed by persons belonging to various socio economic groups, which are strongly influenced by the geographically varying parameters like: living standards, employment opportunities, drug abuse, literacy, migration, type of social activity or type of occupation, natural calamity such as flood, earthquake, drought, tsunami, etc. and inaccessible areas such as forest, mountains, arid areas, etc.

This situation is highly complicated in urban areas in spite of having effective patrol with enhanced security measures. This increases doubts among citizens - how do criminals remain successful in their attempt to escape with huge amount with little trace of their identity?

To understand the nature of crime, to reveal the secrets of criminals and their modus operandi, crime types may be categorised into two. One set of crime involves activities of criminals where they remain successful because of their rigorous practice and training in the target area. This kind of petty crime happens everyday and can happen anywhere (specific to geographical locations) but normally remains confined within buses, railway stations, in public gatherings, overcrowded areas, road side at late nights, isolated places, etc. The loss incurred to the public depends entirely on what the victims have and how careful they are in safeguarding their belongings. It is even difficult to maintain a vigil on such crimes by police force in every nook and corner of neighbourhoods and public roads. This can be controlled or minimised only with public co-operation. On the other hand, police should identify those areas vulnerable for this kind of crime and increase their patrol range. Also, these crimes may be committed by people of any age group who are forced to live below poverty line and are subject to vulnerable socio-economic factors mentioned above.

Criminals and their Knowledge in Geography

In the second category of crime, criminals use geographic understanding in an effective way. It is also known as 'organised crime'. This kind of crime cannot be of daily occurrence, because it requires considerable knowledge about the target area i.e. victims' location, geography of location, understanding on the kind of life people lead, existing vulnerable groups such as children, disabled persons, aged people, period of silence or duration of houses being locked, escape route in times of danger, etc. But once the crime is committed, it can be huge loss in terms of money, human lives or property. These people camouflage themselves in the target area; get involved in local activities to know vulnerable houses, and surrounding neighbourhoods. Soon after obtaining all necessary information and making sure the region is suitable, criminals strike suddenly and vanish from the crime scene.

When criminals have such a mental map of the target area, it is very obvious and necessary that police involved in controlling, identifying and monitoring these activities, should have more sophisticated modern geographic information tool investigative tool capable of supporting huge geographical data to identify hot spot areas. The mental map may also be dependent on whether the killer is a hunter, stalker, or has some other mode of attack, since the type of approach used on a victim also has a relationship to the location of the killer's hoe base.

Rossmo lists the relevant offender styles as:

- Hunter (searches for a specific victim in home territory)
- Poacher (travels away from home for hunting)
- Troller (opportunistic encounters while occupied in other activities)
- Trapper (creates a situation to draw a victim to him)

As the crime activities are region specific and geographically variable, it requires a powerful technology capable of handling both geographic and descriptive data. The integrated approach of these geographic and descriptive data will give clue to the potential areas, possible route of criminals, hiding places, etc. Once mapped, it is possible to adcpt preventive measures to control the increasing crime.

Role of GIS in Combating Crime

GIS being the powerful tool, has the ability to handle vast geographic and descriptive data. The integrated approach of spatial and non-spatial data helps to identify hot spots which can be used to tighten patrol.

The power of criminals is their knowledge of geography of the target area which is set up in-built in their mind from childhood. This crucial factor determines their success rate. When criminals use geographical knowledge by their experience of the target area, then, police should be equipped to handle GIS tools which can become the best resort in nabbing the criminals. Decades of stored crime records can be input into GIS for further analyses. This can reveal some interesting pattern never before noticed by manual analyses.

- Hot spots within city
- Different crime occurrence over a specific area
- Frequency of occurrence, season and time
- Individual crime type
- Pattern of behaviour of certain known criminals released on bail

Also, when police have a spatial data (geographical data) for a considerable period of time, they can do trend analysis in GIS which in turn, help to understand the following:

- Crime-prone areas
- Likely social causes
- Areas requiring community development
- Areas of unemployed youth (social instability)
- Need for patrol

Once the crime records are stored in GIS environment, police can get more details on the following:

- Regular monitoring
- Updating of fresh crimes
- Coincidence of certain crimes would be known only after mapping
- It could reveal certain crime patterns that could not have been imagined before such geographical analysis.

Above all, it helps Government to formulate developmental activities in such a way to enhance their attention in controlling the crime by increasing the patrol and safety measures. It further helps to find out the root cause behind people involved in committing crimes. Using GIS, more analysis is possible which reduce the otherwise laborious task of police in collecting and handling huge geographic data. The following analyses can be performed using GIS:

- Weekly, monthly, quarterly and yearly mapping with reports in tabular form

- Density mapping (Hot Spots)
- Multiple crime type at single location
- Trend analysis using 5-10 years of crime data
- Threshold analysis using historical data to estimate an expected range of crime
- Crime activities within X km radius of commercial, residential, or industrial area
- Vehicle theft location and recovery location to track the movement of criminals

Conclusion

It is understood from the reality that criminals remain successful in their attempts as long as they rely on their mental map in their attempt to target the victims. It is high time that our police force make use of modern geographic tool such as GIS and Global Positioning System (GPS) to increase their probability of catching criminals and monitoring their activities. When police force brings all the crime records into GIS environment, it greatly helps them to understand many hidden links to various crime types.

REFERENCES

Godwin, G.M., (ed). (2000). Criminal Psychology and Forensic Technology: A Collaborative Approach to Effective Profiling. Boca Raton, FL CRC Press.

Rossmo, D.K., (1999) Geographic Profiling, in Jackson, J.L. and Bekerian, D., (eds). Offender Profiling: Theory, Practice and Research, New York: Wiley.

APPLICATIONS OF GEOSPATIAL TECHNOLOGY
A. Ganesh (Ed.), (2006) : pp. 233-251
Satish Serial Publishing House, New Delhi
E-mail: ssp@ndf.vsnl.net.in, hkjain1975@yahoo.com
website: www.satishserial.co

18

Principles of Global Positioning Systems and Its Applications

A. GANESH AND R. NARAYANKUMAR

The Global Positioning System (GPS) is a network of 24 Navstar satellites orbiting at a distance of 20,200 km from the surface of the Earth. Originally established by the U.S. Department of Defense (DOD) at a cost of about US$ 13 billion, access to GPS is free to all users. The system's positioning and timing data are used for a variety of applications, including air, land and sea navigation, vehicle and vessel tracking, surveying and mapping, and asset and natural resource management. With military accuracy restrictions partially lifted in March 1996 and fully lifted in May 2000, GPS can now pinpoint the location of objects as small as a penny anywhere on the Earth surface. GPS is the shortened form of NAVSTAR GPS. This is an acronym for NAVigation System with Time And Ranging.

GPS is a solution for one of man's longest and most troublesome problems. It provides an answer to the question 'Where on earth am I?' One can imagine that this is an easy question to answer. You can easily locate yourself by looking at objects that surround you and position yourself relative to them. But what if you have no objects around you? What if you are in the middle of the desert or in the middle of the ocean? For many centuries, this problem was solved by using the sun and stars to navigate. These methods worked well within certain boundaries. Sun and stars can not be seen when it is cloudy. Also, even with the most precise measurements position cannot be determined very accurately. After the Second World War, it became apparent to the U.S. Department of Defense that a solution had to be found to the problem of accurate and absolute positioning. Several projects and experiments ran during the next 25 years or so. All of these projects allowed positions to be determined but were limited in accuracy or functionality.

At the beginning of the 1970s, a new project was proposed - GPS. This concept promised to fulfill all the requirements of the US government, namely that one should be able to determine ones position accurately, at any point on the Earth's surface, at any time, in any weather conditions. GPS is a satellite-based system that uses a constellation of 24 satellites to give a user an accurate position.

GPS was originally designed for military use at any time, anywhere on the surface of the Earth. Soon after the original proposals were made, it became clear that civilians could also use GPS, and not only for personal positioning. The first two major civilian applications to emerge were marine navigation and surveying. Now-a-days applications range from in car navigation through truck fleet management to automation of construction machinery.

Historical Development of GPS

Satellite-based positioning has been pursued since the 1960's. An early and very successful satellite positioning system was the Navy Navigation Satellite System (TRANSIT). Since its release for commercial use in 1967, the TRANSIT positioning system was often used to determine widely spaced networks covering large regions- even the globe. It was instrumental in establishing modern geocentric datum and in connecting various national datum to a geocentric reference frame. The TRANSIT satellites were orbiting in polar plane at about 1100 km altitude. The TRANSIT satellites were affected more by gravity field variations than the much higher-orbiting GPS Satellites. In addition, their transmissions at 150 and 400 MHz were more susceptible to ionosphere delays and disturbances than the higher GPS frequencies The TRANSIT system was discontinued at the end of 1996 and replaced by GPS. The following is the chronological development of GPS.

System Overview

The total GPS configuration is comprised of three distinct segments

Space Segment	-	Satellites orbiting the earth.
Control Segment	-	Stations positioned on the earth's equator to control the satellites
User Segment	-	Any body that receives and uses the GPS signal

Table 1. Chronological Development of GPS

1978	Launch of first GPS satellite
1982	Prototype Macro meter testing at M.I.T
1983	Geodetic network densification (Eifel, Germany)
	President Regan offers GPS to the world "free of charge"
1984	Geodetic network densification (Montgomery County, Pennsylvania) Engineering survey at Stanford
1985	Precise geoid undulation differences for Eifel network Codeless dual band observations Kinematic GPS Surveying First international symposium on precise positioning with GPS
1986	Challenger accident (January 28) 10 cm aircraft positioning
1987	JPL baseline repeatability test to 0.2-0.04 ppm
1989	Launch of first Block II Satellite Wide area differential GPS (WADGPS) concepts U.S. Coast Guard GPS Information Centre (GPSIC)
1990	GEOID90 for NAD83 datum
1991	NGS ephemeris service
	GIG 91 experiment (Jan 22 - Feb1
1992	Initial solutions to deal with antispoofing (AS)
	Narrow correlator spacing C/A -code
1993	Real-time kinematic GPS
	ACSM ad hoc committee on accuracy standards
	Initial operational capability (IOC) on December 8
	1-2 ppb baseline repeatability
	LAMBDA
1994	IGS service beginning January 1
	Antispoofing implementation (January 31)
	RTCM recommendations on differential GPS (version 2.1)
	National Spatial Reference System Committee (NGS)
	Multiple (single-frequency) receiver experiments for OTF
	Proposal to monitor the earth's atmosphere with GPS (occulations)
1995	Full operational capability (FOC) on July 17
	Precise point positioning (PPP) at JPL
1996	Presidential Decision Directive, first U.S.GPS policy
1998	Announcement on the second GPS civil signal at 1227.60 MHz JPL's automated GPS data analysis service via Internet
1999	Announcement on GPS modernization and third civil GPS signal at 1176.45 MHz
	IGDG (Internet-based global differential GPS) at JPL
2000	Selective availability set to zero
	GPS JPO begins modifications to IIR-M and IIF satellites

Space Segment

The Space Segment of the system consists of the GPS satellites. These space vehicles send radio signals from space. The nominal GPS Operational Constellation consists of 24 satellites orbiting the Earth at approximately 20200 km every 12 hours (Fig. 1). There are often more than 24 operational satellites as new ones are launched to replace older satellites. The satellite orbits repeat almost the same ground track as the earth turns beneath them once each day. The orbit altitude is such that the satellites repeat the same track and configuration over any point approximately each 24 hours (4 minutes earlier each day). There are six orbital planes with nominally four space vehicles in each, equally spaced i.e. 60 degrees apart, and inclined at about fifty-five degrees with respect to the equatorial plane. This constellation provides the user with between five and eight space vehicles visible from any point on the Earth.

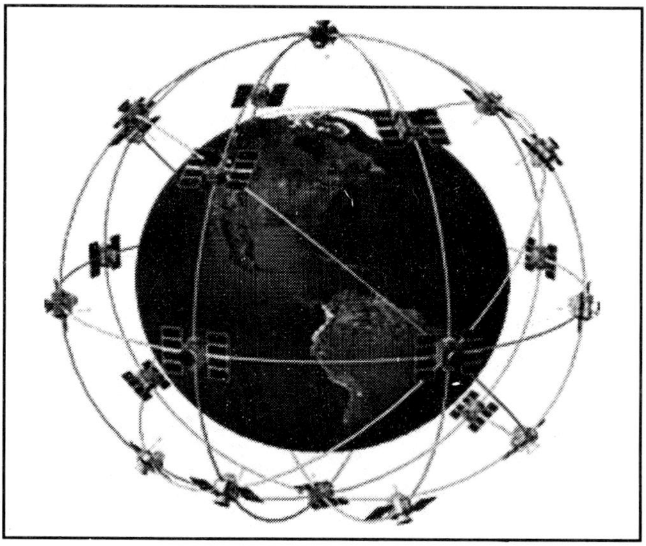

Fig.1 GPS Satellite Constellation

Each GPS satellite has several very accurate atomic clocks on board. The clocks operate at a fundamental frequency of 10.23 MHz. This is used to generate the signals that are broadcast from the satellite (Fig. 2). The satellites broadcast two carrier waves constantly. These carrier waves are in the L-Band (used for radio), and travel to earth at the speed of light. These carrier waves are derived from the fundamental frequency, generated by a very precise atomic clock.

L1 carrier is broadcast at 1575.42 MHz (10.23 x 154)

L2 carrier is broadcast at 1227.60 MHz (10.23 x 120).

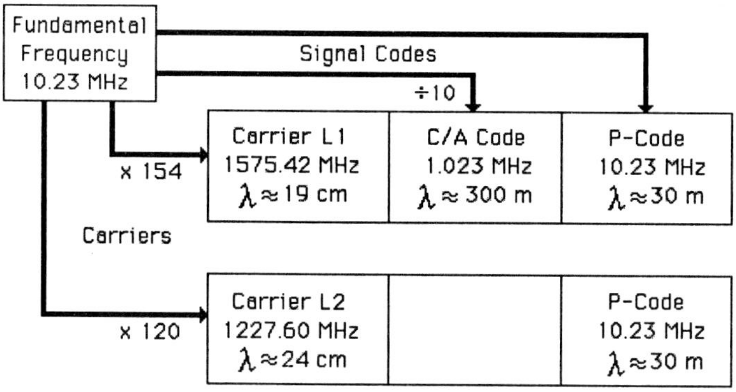

Fig. 2 GPS Signal Structure

The L1 carrier then has two codes modulated upon it. The C/A Code or Coarse/Acquisition Code is modulated at 1.023MHz (10.23/10) and the P-code or Precision Code is modulated at 10.23MHz. The L2 carrier has just one code modulated upon it. The L2 P-code is modulated at 10.23 MHz. GPS receivers use the different codes to distinguish between satellites. The codes can also be used as a basis for making pseudo range measurements and therefore calculate a position.

Control Segment

The Control Segment consists of a system of tracking stations located around the world. The Control Segment consists of one master control station, five monitor stations and four ground antennas distributed amongst five locations roughly on the Earth's equator (Fig. 3). The Control Segment tracks the GPS satellites, updates their orbiting position and calibrates and synchronizes their clocks. A further important function is to determine the orbit of each satellite and predict it's path for the following 24 hours. This information is uploaded to each satellite and subsequently broadcast from it. This enables the GPS receiver to know where each satellite can be expected to be found. The satellite signals are read at Ascension, Diego Garcia and Kwajalein. The measurements are then sent to the Master Control Station in Colorado Springs where they are processed to determine any errors in each satellite. The information is then sent back to the four monitor stations equipped with ground antennas and uploaded to the satellites.

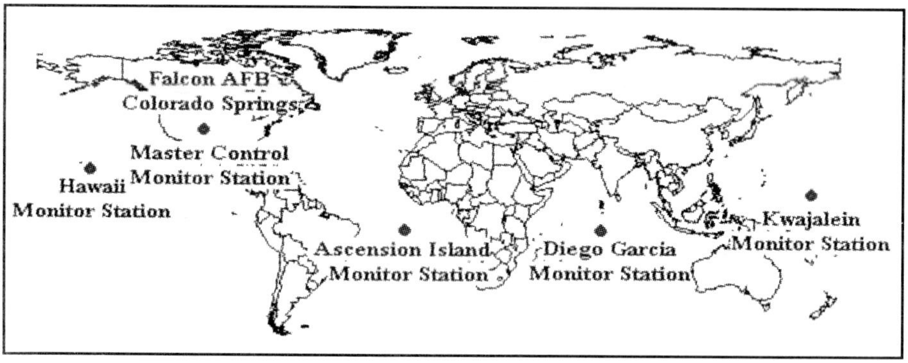

Fig.3 Network of GPS Master Control and Monitoring Stations

User Segment

The User Segment comprises of anyone using a GPS receiver to receive the GPS signal and determine their position and/or time. Typical applications within the user segment are land navigation for hikers, vehicle location, surveying, marine navigation, aerial navigation, machine control etc.

Principles of GPS

Resection

All GPS positions are based on measuring the distance from the satellites to the GPS receiver on the Earth. This distance to each satellite can be determined by the GPS receiver. If you know the distance to three points relative to your own position, you can determine your own position relative to those three points. From the distance to one satellite we know that the position of the receiver must be at some point on the surface of an imaginary sphere which has it's origin at the satellite. By intersecting three imaginary spheres the receiver position can be determined (Fig. 4).

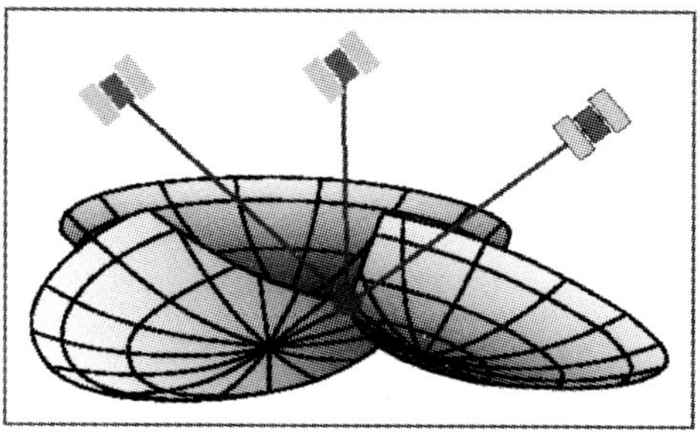

Fig.4 Intersection of Three Imaginary Spheres

Four unknowns

The problem with GPS is that only pseudo ranges and the time at which the signal arrived at the receiver can be determined. Thus there are four unknowns to determine, position (X, Y, Z) and time of travel of the signal. Observing to four satellites produces four equations which can be solved, enabling these unknowns to be determined.

Calculating the Distance to the Satellite

In order to calculate the distance to each satellite, one of Isaac Newton's laws of motion is used:

Distance = Velocity x Time

GPS requires the receiver to calculate the distance from the receiver to the satellite. The Velocity is the velocity of the radio signal. Radio waves travel at the speed of light 290,000 km per second (186,000 miles per second).

Time is the time taken for the radio signal to travel from the satellite to the GPS receiver. This is a little harder to calculate, since you need to know when the radio signal left the satellite and when it reached the receiver.

Calculating Time

The satellite signal has two codes modulated upon it, the C/A code and the P-code. The C/A code is based upon the time given by a very accurate atomic clock. The receiver also contains a clock that is used to generate a matching C/A code. The GPS receiver is then able to 'match' or correlate the incoming satellite code to the receiver generated code.

The C/A code is a digital code that is 'pseudo random' or appears to be random. In actual fact it is not random and repeats one thousand times every second. In this way, the time taken for the radio signal to travel from the satellite to the GPS receiver is calculated.

Errors and Precision

Up until this point, it has been assumed that the position derived from GPS is very accurate and free of error, but there are several sources of error that degrade the GPS position from a theoretical few meters to tens of meters. These error sources are: Ionospheric and atmospheric delays, Satellite and Receiver Clock Errors, Multipath, Dilution of Precision, Selective Availability (S/A) and Anti Spoofing (A-S)

Ionospheric and atmospheric delays

As the satellite signal passes through the ionosphere, it can be slowed down, the effect being similar to light refracted through a glass block. These atmospheric delays can introduce an error in the range calculation as the velocity of the signal is affected. Light only has a constant velocity in a vacuum. The ionosphere does not introduce a constant delay on the signal. There are several factors that influence the amount of delay caused by the ionosphere. Satellite elevation-Signals from low elevation satellites will be affected more than signals from higher elevation satellites. This is due to the increased distance that the signal passes through the atmosphere. The density of the ionosphere is affected by the sun. At night, there is very little ionospheric influence. In the day, the sun increases the effect of the ionosphere and slows down the signal. Water vapour also affects the GPS signal. Water vapor contained in the atmosphere can also affect the GPS signal. This effect, which can result in a position degradation can be reduced by using atmospheric models.

Satellite and Receiver Clock Errors

Even though the clocks in the satellite are very accurate to about 3 nanoseconds, they do sometimes drift slightly and cause small errors, affecting the accuracy of the position. The US Department of Defense monitors the satellite clocks using the Control Segment and can correct any drift that is found.

Multipath

Multi path occurs when the receiver antenna is positioned close to a large reflecting surface such as a lake or building. The satellite signal does not travel directly to the antenna but hits the nearby object first and is reflected into the antenna creating a false measurement. Multipath can be reduced by the use of special GPS antennas that incorporate a ground plane a circular, metallic disk about 50cm in diameter that prevent low elevation signals reaching the antenna. For highest accuracy, the preferred solution is use of a choke ring antenna. A choke ring antenna has four or five concentric rings around the antenna that trap any indirect signals. Multi path only affects high accuracy, survey type measurements. Simple handheld navigation receivers do not employ such techniques.

Dilution of Precision

The Dilution of Precision (DOP) is a measure of the strength of satellite geometry and is related to the spacing and position of the satellites in the sky. The DOP can magnify the effect of satellite ranging errors. Different types of Dilution of Precision or DOP can be calculated depending on the dimension. VDOP (Vertical Dilution of Precision) gives accuracy degradation in vertical direction. HDOP (Horizontal Dilution of Precision) gives accuracy degradation in horizontal direction. PDOP (Positional Dilution of Precision) gives accuracy degradation in 3D position. GDOP (Geometric Dilution of Precision) gives accuracy degradation in 3D position and time. The most useful DOP to know is GDOP since this is a combination of all the factors.

Selective Availability (S/A)

Selective Availability is a process applied by the U.S. Department of Defense to the GPS signal. This is intended to deny civilian and hostile foreign powers the full accuracy of GPS by subjecting the satellite clocks to a process known as 'dithering' which alters their time slightly. It is worth noting that S/A affects civilian users using a single GPS receiver to obtain an autonomous position. Users of differential systems are not significantly affected by S/A.

Anti Spoofing (A-S)

Anti-spoofing is similar to S/A ,it's intention is to deny civilian and hostile powers access to the P-code part of the GPS signal and hence force use of the C/A code which has S/A applied to it.

Blunders

Blunders can result in errors of hundred of kilometers. It may be due to control segment mistakes due to computer or human error can cause errors from one meter to hundreds of kilometers. User mistakes, including incorrect geodetic datum selection, can cause errors from 1 to hundreds of meters. Receiver errors from software or hardware failures can cause blunder errors of any size.

Factors Affecting Accuracy

The following are the factors that affect the GPS survey accuracy. They are also the main GPS configuration parameters.

a. *Number of Visible Satellites (SV)*: If you require 3-dimensional coordinates (northerning, easting, and elevation), a minimum of four satellites is needed. If you require 2-dimensional coordinates, at least three satellites are needed, but the 2 dimensional mode can significantly reduce the final accuracy.

b. *Position Dilution of Precision (PDOP)*: This is a measure of the current satellite geometry. It takes into account each satellite's location relative to the other satellites in the constellation. A low DOP indicates a higher probability of accuracy. A high DOP indicates a lower probability of accuracy. A PDOP of 4 or below gives excellent positions. A PDOP between 5 and 8 is acceptable. A PDOP of 9 or more is poor.

c. *Satellite Elevations*: When a satellite is low on the horizon, the satellite signals must travel a greater distance through the atmosphere, resulting in a lower signal strength and delayed reception by the GPS receiver. Position data should be collected using only satellites that are at least 15° above the horizon.

d. *Multipath*: Satellite signals can reflect off larger nearby objects, such as buildings or cars, causing the GPS antenna to receive an erroneous signal (longer path length = false location). This phenomenon is known as multipath, which can induce errors of dozens of meters. Optimal accuracy can be obtained by collecting data in an environment devoid of large reflective surfaces and that has a clear view of the sky. Avoid standing next to a building.

e. *Distance between Base Station and Rover Receivers*: Accuracy degrades as the distance between base station and rover increases. To obtain sub meter accuracy, you can generally only capture data within 300 m of a base station. Using the satellite Differential GPS services (like Landstar, the subscription service for GIS Lab), this is not an issue-all of North America is within the baseline length limit for the network of bases stations.

f. *Signal-to-Noise Ratio (SNR)*: This is a measure of the information content of a signal relative to the signal's noise. As this proportion decreases, information gets lost in the noise. A value above 20 is very good. The quality of a position is degraded if the signal strength of any satellite in the constellation is below 6. The default SNR mask is 6. The satellite signals do not penetrate metal surfaces, buildings, tree trunks, or similar objects. When recording tree location data, for example, consider placing the antenna a meter or two away from the tree.

g. Occupation Time at a Point: Accuracy at a spot can be improved by collecting over a longer period of time at the location, then averaging these positions. However, unless you are willing to occupy a point for half a day or so, the sample will not be big enough to be statistically better than collecting 3 positions (~3 seconds) and moving on to the next point.

h. Differential Correction: This involves using a second GPS receiver at a known location (a base station) to correct for the drift of GPS signals.

Differentially Corrected Positions [DGPS]

Many of the errors affecting the measurement of satellite range can be completely eliminated or at least significantly reduced using differential measurement techniques. DGPS allows the civilian user to increase position accuracy from 100 m to 2-3 m or less, making it more useful for many civilian applications. Differential GPS (DGPS) is a means of correcting for some system errors by using the errors observed at a known location to correct the readings of a roving receiver. The basic concept is that the reference station 'knows' its position, and determines the difference between that known position and the position as determined by a GPS receiver. This error measurement is then passed to the roving receiver which can adjust its indicated position to compensate. Unfortunately, the error depends on the particular satellites used to compute the position, so the reference station can't just say 'move all positions 100 metres south'. The differential reference station computes the errors in the pseudo range measurements for each satellite in view separately, and broadcasts the error information, and other system status information, by some means. A differential beacon receiver receives and decodes this information, and sends it to the "differential ready" GPS receiver. The GPS receiver combines this information with the individual pseudo range measurements it makes, before calculating the position. DGPS will eliminate the error introduced by Selective Availability, and errors caused by variations in the ionosphere, resulting in reported positions within about 10 metres of the true position 95 per cent of the time, for typical marine DGPS systems using inexpensive navigation receivers. Better receivers can get within three meters, or so. The DGPS correction data can be used as far as 1500 KM from the reference station depending on the DGPS setup - if the DGPS is part of a larger monitoring network. A typical DGPS involves two types of receivers- the reference and rover receivers, using them the Differentially Corrected Positions are found out.

Reference Receiver

The Reference receiver antenna is mounted on a previously measured point with known coordinates. The receiver that is set at this point is known as the Reference Receiver or Base Station. The receiver is switched on and begins to track satellites. It can calculate an autonomous position. Because it is on a known point, the reference receiver can estimate very precisely what the ranges to the various satellites should be. The reference receiver can, therefore, work out the difference between the computed and measured range values. These differences are known as corrections. The reference receiver is usually attached to a radio data link which is used to broadcast these corrections.

Rover receiver

The rover receiver is on the other end of these corrections. The rover receiver has a radio data link attached to it that enables it to receive the range corrections broadcast by the Reference Receiver. The Rover Receiver also calculates ranges to the satellites. It then applies the range corrections received from the Reference. This lets it calculate a much more accurate position than would be possible if the uncorrected range measurements were used. Using this technique, all of the error sources are minimized, hence the more accurate position. It is also worthwhile to note that multiple Rover Receivers can receive corrections from one single Reference.

Surveying with GPS

Probably even more important to the surveyor or engineer than the theory behind GPS, are the practicalities of the effective use of GPS. Like any tool, GPS is only as good as it's operator. Proper planning and preparation are essential ingredients of a successful survey, as well as an awareness of the capabilities and limitations of GPS.

Advantages using GPS

GPS has numerous advantages over traditional surveying methods:

1. Inter visibility between points is not required.
2. Can be used at any time of the day or night and in any weather.
3. Produces results with very high geodetic accuracy.
4. More work can be accomplished in less time with fewer people.

Limitations of GPS

In order to operate with GPS it is important that the GPS Antenna has a clear view to at least four satellites. Sometimes, the satellite signals can be blocked by tall buildings, trees etc. Hence, GPS cannot be used indoors. It is also difficult to use GPS in town centers or woodland. Due to this limitation, it may prove more cost effective in some survey applications to use an optical total station or combine use of such an instrument with GPS.

GPS Measuring Techniques

GPS survey systems are one of the first uses of commercial GPS. These units are more accurate than the typical navigation units but rely on post-processing of the data collected by roving receivers and a fixed reference receiver, and on averaging the data collected over a period of time to get the increased accuracy. These systems can have an accuracy of better than 1 cm for the very expensive models.

There are several measuring techniques that can be used by most GPS Survey Receivers. The surveyor should choose the appropriate technique for the application.

Static: Used for long lines, geodetic networks, tectonic plate studies etc. Offers high accuracy over long distances but is comparatively slow. Two GPS receivers are required for this survey and they can be placed in separate locations for a period of time for short distances 2 mins up to 1 hr and the raw pseudo range data can be collected. This data can then be post-processed and a baseline established i.e. range and bearing. The position can be as good as 1 mm but may be less good under poor satellite geometry or larger distances. This method can then be used to transfer the knowledge of one accurate point to a new point.

Rapid Static: Used for establishing local control networks, network densification etc. Offers high accuracy on baselines up to about 20 km and is much faster than the Static technique.

Kinematic: Used for detail surveys and measuring many points in quick succession.

Very efficient way of measuring many points that are close together. However, if there are obstructions to the sky such as bridges, trees, tall buildings etc., and less than four satellites are tracked, the equipment must be reinitialized which can take 5-10 minutes. A processing technique known as On the-Fly (OTF) has gone a long way to minimize this restriction.

RTK: RTK systems work in similar ways to DGPS short range systems but mathematically like kinematic and can achieve accuracies of around 7 cm in real time. Real Time Kinematic uses a radio data link to transmit satellite data from the Reference to the Rover. This enables coordinates to be calculated and displayed in real time, as the survey is being carried out. Used for similar applications as kinematic. This technique is however reliant upon a radio link, which is subject to interference from other radio sources and also line of sight blockage.

Uses of GPS Technology

GPS technology has matured into a resource that goes far beyond its original design goals. These days' people from a variety of professions are using GPS in ways that make their work more productive, safer, and sometimes even easier. There are five main uses of GPS today:

1.Location- determining a basic position.

2.Navigation - getting from one location to another.

3.Tracking - monitoring the movement of people and things.

4.Mappin g- creating maps.

5.Timing - providing precise timing.

Location

The first and most obvious application of any Location Based Service such as GPS is the simple determination of a 'position' or location. GPS was the first positioning system to offer highly precise location data for any point on the planet, in any weather. Knowing the precise location of something, or someone, is especially critical when the consequences of inaccurate data are measured in human terms.

Navigation

GPS helps you determine exactly where you are, but sometimes it is more necessary to know how to get somewhere else. Recall that GPS was originally designed to provide navigation information for ships and planes. So it's no surprise that while this technology is appropriate for navigating on water, it's also very useful in the air and on the land.

Tracking

GPS used in conjunction with communication links and computers can provide the backbone for systems tailored to applications in agriculture, mass transit, urban delivery, public safety, and vessel and vehicle tracking. Therefore, more and more police, ambulance, and fire departments are adopting systems like GPS-based AVL (Automatic Vehicle Location) Manager to pinpoint both the location of the emergency and the location of the nearest response vehicle on a computer map. With this kind of clear visual picture of the situation, dispatchers can react immediately and effectively.

Mapping

Mapping the planet has never been an easy task, but GPS today is being used to survey and map it precisely, saving time and money in this most stringent of all applications. GPS can help generate maps and models of everything in the world, mountains, sea, rivers, cities and help manage endangered animals, archaeological treasures, precious minerals and all sorts of resources, as well as accurately managing the effect of damage and disasters. For example peace brokers used GPS generated maps to determine the partitions of Bosnia under the Dayton Peace Accord, and GPS was used to map Cambodia accurately for the UN stabilisation force introduced after the civil war ended in 1993. In fact large tracks of Cambodia had never been mapped at all, due to inhospitable terrain and low population density. GPS solved this in weeks, whereas it would have taken months prior to the introduction of GPS.

Timing

GPS can also be used to determine precise time, time intervals, and frequency. There are three fundamental ways we use time:

- As a universal marker,
- As a way to synchronize
- To provide an accurate, unambiguous sense of duration.

As discussed before, GPS satellites carry highly accurate atomic clocks. And in order for the system to work, GPS receivers here on the ground synchronize themselves to these clocks. That means that every GPS receiver is, in essence, an atomic accuracy clock. Astronomers, power companies, computer networks, communications systems, banks, and radio and television stations can benefit from this precise timing.

Competitors to GPS

As it stands at the moment there are two satellite systems being marketed as current and future satellite systems billed as being able to provide location-based services.

Glonass

GLONASS is the Russian Federation's satellite navigation system. In many ways it is very similar to GPS. There are 24 satellites in the full constellation; each satellite transmits various data on two L-band carriers; there is one navigation signal that has been authorized for civilian use, and further navigation signals that have been reserved solely for Russian military use; there is a ground segment that monitors and control the satellites; and users passively receive its signals and are able to navigate with accuracy's of few tens of metres or better. However, it is too easy from a user's perspective to presume that GLONASS is the same as GPS. Rather, there are some key differences between the two systems. In many ways, GLONASS is a more elegant and economical system design than GPS. Unfortunately there are a number of quality control issues and much more serious funding problems. The GLONASS system was declared fully operational on 18 January 1996. Since then there have been few occasions when the full complement of 24 satellites was operating. In addition to satellite outages eight satellites have been withdrawn from service since this date. So by late 1997 only 15 satellites were available for navigation.

Recent announcements by Russian officials may mean that 6 years of neglect of the GLONASS constellation of satellites will stop. Solid government financial commitments have been obtained to return the system to fully operational status by 2006, Russian officials say. Also, this new generation of GLONASS satellites (Glonass-M) will have longer in-orbit lives that should ease pressure on the system, they said. They are designed to have a seven-year service life, compared to three years for the current GLONASS. Whether this financial commitment will materialize is another thing, but even in its current weakened state, GLONASS still has potential as a stand-alone navigation system, and as an augmentation to GPS.

Galileo

The Galileo satellite radio navigation system is an initiative launched by the European Union and the European Space Agency

(ESA).The project architects plan deployment in 2006-7, becoming operational in 2008 at a yearly cost of €220 m. It is projected by 2008 a constellation of 30 satellites should be available. The technology behind Galileo is designed to be more accurate and more reliable than GPS or GLONASS. This will allow safety-critical systems - such as air traffic control, and ship and car navigation - to be run on the technology. The system should also guarantee coverage to previously inaccessible areas such as those that are either blocked by buildings or isolated areas at high latitudes.

However, it is estimated that the Galileo project could cost more than €3bn ($2.6 bn). With some of the member countries facing recession, now might not be the time to fund the idea. Britain and the Netherlands have already questioned the timing, and Germany has raised issues over costs. In contrast, France and Italy whose aerospace industries both stand to benefit from the system are looking forward to a go-ahead. The decision to push ahead with the project remains up in the air for now.

REFERENCES

Aerospace, (1968). The 1968 Aerospace Year Book. Washington, D.C.: Books, Inc.

Agrawal, N. K., (2004). Essentials of GPS. Hydrabad: Spatial Networks Pvt. Ltd.

Buckeye, K.R., (2004). GIS - GPS for Distance - Based User Fee System, GIS Development, 8 (11), pp. 34-37.

Burchell, Wade, Lt. Cmdr, (1997). Reserve Helos Get Magic Lantern Naval Institute Proceedings, 74-75

Curtis Peebles, (1987). Guardians: Strategic Reconnaissance Satellites, Novato, CA: Presidio.

Galileo (2005). Galileo Home Page at http:\\Galileo-pgm.org

Ganesh, A. and Narayanakumar R., (2003). Principles of GPS. Tiruchirappalli: Bharathidasan University.

Gartner, George, V. Radoczky and G. Retscher, (2005). Location Technologies for Pedestrian Navigation, GIS Development, 9 (4), pp. 22-31.

GLONASS, (2005) GLONASS Home Page at http:\\rssi.ru/SFCSIC/SFCSIC_main.html.

Gouzhva, Yuri G. and Arvid G. Guevorkyan, Arkady B. Bassevich, Pyotr P. Bogdanov, (1992). High-Precision Time and Frequency Dissemination With Glonass, GPS World, 3, (7), pp. 40-49

GPS, (2005). GPS Home Page at US Coast Guard www.navcen.useg.gov./default.htm

GPS World, (1992). Protesters invade Rockwell Plant, Damage GPS Satellite, GPS World, 3 (6), p. 12

GPS World, (1992). Satellite Set Healthy, Launch Planned for Mid-June, GPS World, 3 (6), p.12

GPS World, (1993). FAA Approves GPS for IFR Operations, GPS World, 4, (7), p.18

He, Xiufeng, W. Sang, Y. Chen and X. Ding (2005) Steep – Slope Monitoring. GPS Multiple Antenna System at Xiaowan Dam, GPS World, 16 (11), pp. 20-25.

Higbie, Paul R. and Norman K. Blocker. (1994). Detecting Nuclear Detonations with GPS, GPS World, 5 (2), pp.48-50

Hofmann-Wellenhof, B. Lichtenegger, H and Collins.J, (2001). GPS Theory and Practice, New York: Springer Wien.

Ivan A. Getting. (1993). The Global Positioning System, IEEE Spectrum, pp.36-47

Jethwani, Hiresh, (2004). GPS Navigation in Indian Automobiles, GIS Development, 8 (11) p. 44.

Johnson, Nicolas, L., (1994). GLONASS Spacecraft, GPS World, 5 (11), pp. 51-58

Leica (2003). GPS Basics: Introduction to GPS, Leica Version 1.0.

Leick Alfred, (2004). GPS Satellite Surveying, Third Edition, New Jersey: John Wiley and Sons Inc.

Kaplan E.D. (Ed). (1996). Undernawy GPS Principles an Application. Norwood. MA: Artech House.

Kennedy, Michael (2002). The Global Positioning System and GIS : An Introduction, London : Taylor and Francis

Kraak, Menno-Jan and Allan Brown, (2001). Web Cartography. London : Taylor and Francis.

Parkinson, Bradford, W., (1994). GPS Eyewitness: The Early Years. GPS World, 5 (9), pp. 32-45.

Schiestl, Andrew J., (2004). GPS and Alarka Rail Safety. GPS World, 15 (3), pp. 14-19.

Stine, G. Harry (1971). ICBM. New York: Orion Books.

Syed, Salmand and E. Cannon (2005). Map-Added GPS Navigation, GPS World, 16 (11), pp. 39-44.

Tsui, James Bao-Yen, (2005). Fundamentals of Global Positioning System Receivers. A Software Approach, Texas: Wiley.

Whitford, Marty, (2005) Hurricane Hunters : GPS Dropsondes Trace Katrina's Course, GPS World, 16 (11), pp. 20-22.

Worthman, Ernest, (1997). Global Positioning Systems - Technology into the 21st Century, RF Design, pp. 44-54

Wysocki, Joseph, Lt. Col., (1991). GPS and Selective Availability - The Military Perspective. GPS World, 2, (7), pp. 38-44

Xie, Yicheen, D. Goby, R. Raymond, R. Pizzi, (2004). Going Beyard Automatic Vehicle Location, GIS Development, 8 (11), pp. 30-33.

Web Resources

http://www.aero.org/education/index.html

http://www.buzzle.com/editorials/9-10-2004-115.asp?viewPage=2

http://www.colorado.edu/geography/gcraft/contents.html

http://www.colorado.edu/engineering/ASEN/asen5090/GPS3.sm.jpg

http://igscb.jpl.nasa.gov/overview.html

http://www.gmat.unsw.edu.au/snap/gps/gps_survey/chap2/211.htm

http://www.gmat.unsw.edu.au/snap/gps/gps_survey/chap3/311.htm

http://www.gps-practice-and-fun.com/personal-location-tracking.html

http://hyperphysics.phy-astr.gsu.edu/hbase/gpsrec.html

http://www.montana.edu/places/gps/understd.html

http://www.navcen.uscg.mil/loran/loranff.htm

http://www.navcen.uscg.mil/omega/omegaff.htm

http://www.navcen.uscg.mil/gps/gpsff.htm

http://www.palowireless.com

http://science.howstuffworks.com/nuclear2.htm

http://www.trimble.com/gps/how.html

APPLICATIONS OF GEOSPATIAL TECHNOLOGY
A. Ganesh (Ed.), (2006) : pp. 253-266
Satish Serial Publishing House, New Delhi
E-mail: ssp@ndf.vsnl.net.in, hkjain1975@yahoo.com
website: www.satishserial.co

19

Integration of GPS and GIS

R. NARAYANAKUMAR AND A. GANESH

GIS, an acronym of Geographical Information System, is an information system used for management decisions and allows one to create, maintain and query electronic databases of information normally displayed on maps. These databases are spatially-oriented, the fundamental integrating element being the position on the Earth's surface. GIS consists of a set of computerized tools and procedures that can be used to effectively encode, store, retrieve, overlay, correlate, manipulate, analyze, query, display (both graphically and numerically) and disseminate Earth related information. The main difference or factor which separates GIS from other information storage and retrieval systems is the spatial and specific vocational features given in a co-ordinate form, for reference and use as an important variable in quantitative analysis.

The spatial data is generally in the form of maps, which could be showing topography, geology, soil types, forest and vegetation, land use / land cover, water resources, GPS reading, etc., and stored as layers in a digital form in the computer. Besides spatial data, one could also have attribute data, like statistics, texts and tables. New maps can be easily generated by integrating many layers of data in a computer. Thus, a GIS has a database of multiple information layers that can be manipulated or analyzed to evaluate relationships among the selected elements, from all that is available.

The power of GIS lies in its ability to identify relationships between features based on their locations and their attributes. GIS allows one to view these relationships in maps, charts, tabular forms etc. In GIS one has the ability to create different layers of data. For example, one can create layers for roads, traffic signs, accidents, buildings, fire hydrants, water lines, etc and once the layers are created one can superimpose different layers and analyze the relationship between various features and their attributes. In order to analyze the data one requires relevant and current information, which can be obtained from various sources. Some data can be obtained from existing sources e.g., population, census, roads etc. However, other data specific to a project may not be readily available and may have to be collected for a specific project with GPS. Once the data is collected in GPS receivers, it can be directly transferred to a GIS System. This data will have the spatial location as well as feature/attribute data attached to it.

Integrating – GPS/GIS Technology

Combining the GPS data with GIS allows for greater capabilities than what GPS and GIS can provide individually. With the combination of two technologies one is able to display the 'FIELD/ACTUAL SITE' on a PC and make informed decisions. There is no need to make specific site visits or review several documents/drawings. Also, another benefit of the integration is the fact that the data can be shared by unlimited users in various departments for their own specific needs and analysis.

Another important advance in this technology has been the introduction of software which allows bringing into GIS not only GPS position information but a digital picture (satellite image) of the same area. With the software, one can study relationships between features and also view actual photographs of the features on the PC.

Pitfalls of Integration

Combination of GPS/GIS technology is limited by the amount of data. As explained earlier one needs lot of good data for conducting analysis. While some data is available, a lot of data has to be generated by the users for their use. Sometimes collection of field data could turn out to be time consuming and expensive.

Data collected must be accurate and meet the correct formats. For example one need to make sure that all the "layers" of data displayed are in same units (feet/meters) and the projections and datums match. Without this the analysis will not be correct. However, the integration of GPS with GIS brings the real world to the desktop. What could take days to visit a specific site and analyze can now be performed on your desktop.

Process of Integration

Once GPS readings were available they need to be processed further before being converted into desired format such as maps. Both GPS and GIS are system executed through software, all the processes of integration have to flow through them. There is a variety of GIS software available in the market, viz., ArcMap, ArcView, Geomedia, etc. Some GPS instruments like the products from Leica are executed through the software Arc Pad - A Product of ESRI. The processes of integration thus involve the modules of both GPS and GIS software. The following are the major processes involved in the integration.

1. Editing and Validation
2. Transformation
3. Projection
4. Mapping and Map Layout

Editing and Validation

There are generally two different types of errors could be anticipated from a GPS reading, viz., errors related to the position of the point, errors related to linear feature.

Point positioning errors may be due to varied reasons, discussed elsewhere in this book. These errors related to points could be solved by precise positioning techniques, observing the same point at different time periods and satellite availability.

Errors related to linear features are of two types, viz., over shoot and Undershoot. Overshoot is the line gone beyond its target snap position. While Undershoot is related to the line which is felt short of some distance before its target snaps position. Overshoot could be corrected by trimming the extended line while the undershoot could be snapped to the target snap position by extending the line segment.

Transformations

The purpose of a transformation is to transform coordinates from one system to another. There are several different transformation approaches exist. The choice of the transformation approach lies in the accuracy requirement (Fig.1). However, the procedure for the determination of transformation parameters is same for all approaches.

To do transformation, coordinates must be available in both coordinate systems. That is in WGS 94 and in the local system. At least

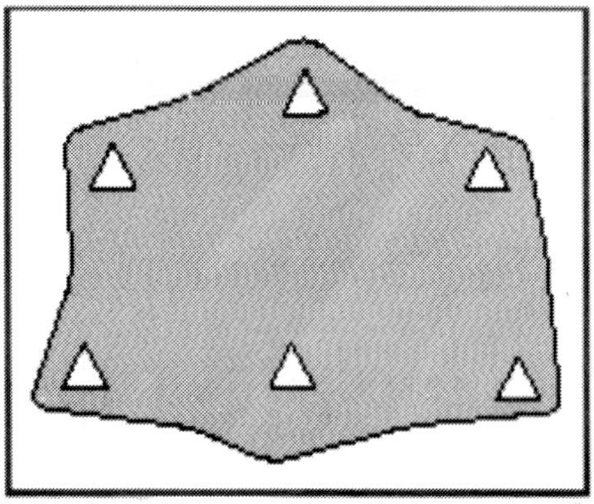

Transformations apply within an area of common points

Fig.1 Determination of Transformations

three, preferably four common points should have been observed. The more common points included the more will be the precision. Common points are point for which coordinates and Orthometric heights are known in the local system. It is important to note that the transformation will apply to points in the area bound by the common points. Points outside of this area should not be transformed using the calculated parameters but should form part of a new transformation area. It is common for GPS surveys to be measured to a much higher accuracy than older surveys measured with traditional optical instruments. In the vast majority of cases, the previously measured points will not be as accurate as the new points measured with GPS. This may create some times non-homogeneity in the network.

In order to transform a coordinate from one system to another, the origins and axes of the ellipsoid must be known relative to each other. From this information, the shift in space in X, Y and Z from one origin to the other can be determined. The following are a few important transformation approaches:

a. *T hree-parameter methods*: The simplest datum transformation method is a geocentric, or three-parameter, transformation. The geocentric transformation models the differences between two datums in the X,Y,Z coordinate system. One datum is defined with its center at 0,0,0. The center of the other datum is defined at some distance (DX,DY,DZ) in meters away (Fig. 2 and 3).

Usually the transformation parameters are defined as going "from" a local datum "to" WGS 1984 or another geocentric datum.

b. *Seven-parameter methods*: A more complex and accurate datum transformation is possible by adding four more parameters to a geocentric transformation. The seven parameters are three linear shifts (DX,DY,DZ), three angular rotations around each axis (rx,ry,rz), and scale factors (Fig. 4).

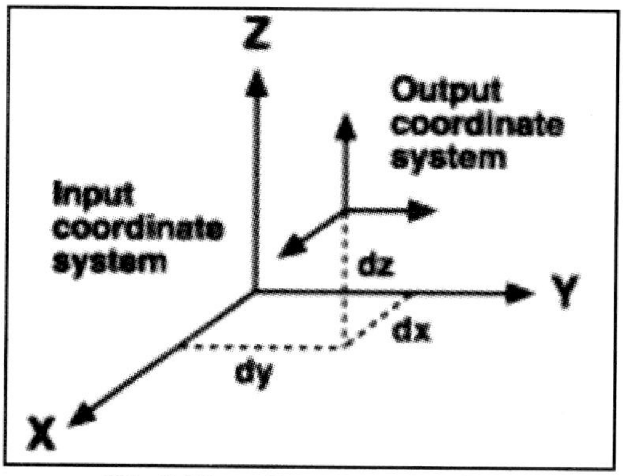

Fig.2 Three Parameter Method

$$\begin{bmatrix} X \\ Y \\ Z \end{bmatrix}_{new} = \begin{bmatrix} \Delta X \\ \Delta Y \\ \Delta Z \end{bmatrix} + \begin{bmatrix} X \\ Y \\ Z \end{bmatrix}_{original}$$

Fig.3 Three Parameter Method

$$\begin{bmatrix} X \\ Y \\ Z \end{bmatrix}_{new} = \begin{bmatrix} \Delta X \\ \Delta Y \\ \Delta Z \end{bmatrix} + (1+s) \cdot \begin{bmatrix} 1 & r_z & -r_y \\ -r_z & 1 & r_x \\ r_y & -r_x & 1 \end{bmatrix} \cdot \begin{bmatrix} X \\ Y \\ Z \end{bmatrix}_{original}$$

Fig.4 Seven Parameter Method

The rotation values are given in decimal seconds, while the scale factor is in parts per million (ppm). The rotation values are defined in two different ways. It's possible to define the rotation angles as positive either clockwise or counterclockwise as you look toward the origin of the X,Y,Z systems (Fig. 5).

FThe previous equation is called the Coordinate Frame Rotation transformation. The rotations are positive counterclockwise. Europe uses a different convention called the Position Vector transformation. Both methods are sometimes referred to as the Bursa-Wolf method.

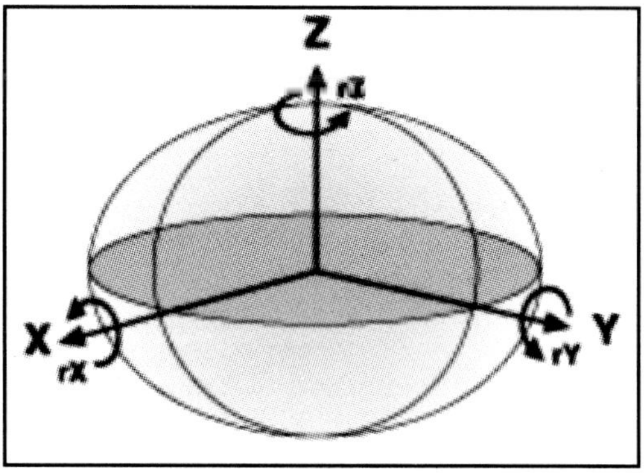

Fig.5 Rotation Angles

Molodensky Method:- The Molodensky-Badekas method is a variation of the seven-parameter methods. It has an additional three parameters that define the XYZ origin of rotation. Sometimes this point is known as the origin of the datum, or geographic coordinate system. Given the XYZ origin of rotation point, it is possible to calculate an equivalent Coordinate Frame transformation. The DX, DY, and DZ values will change but the rotation and scale values will remain the same.

The Molodensky method converts directly between two geographic coordinate systems without actually converting to an X,Y,Z system. The Molodensky method requires three shifts (DX,DY,DZ) and the differences between the semi major axes (Da) and the flattening (Df) of the two spheroids. The Projection Engine automatically calculates the spheroid differences according to the datums involved.

$$(M+h)\Delta\varphi = -\sin\varphi\cos\lambda\Delta X - \sin\varphi\sin\lambda\Delta Y + \cos\varphi\Delta Z + \frac{e^2\sin\varphi\cos\varphi}{(1-e^2\sin^2\varphi)^{1/2}}\Delta a + \sin\varphi\cos\varphi(M\frac{a}{b}+N\frac{b}{a})\Delta f$$

$$(N+h)\cos\varphi\Delta\lambda = -\sin\lambda\Delta X + \cos\lambda\Delta Y$$

$$\Delta h = \cos\varphi\cos\lambda\,\Delta X + \cos\varphi\sin\lambda\,\Delta Y + \sin\varphi\,\Delta Z - (1-e^2\sin^2\varphi)^{1/2}\Delta a + \frac{a(1-f)}{(1-e^2\sin^2\varphi)^{1/2}}\sin^2\varphi\,\Delta f$$

h ellipsoid height (meters)
j latitude
l longitude
a semi major axis of the spheroid (meters)
b semi minor axis of the spheroid (meters)
f flattening of the spheroid
e eccentricity of the spheroid

M and N are the meridional and prime vertical radii of curvature, respectively, at given latitude. The equations for M and N are

$$M = \frac{a(1-e^2)}{(1-e^2\sin^2\varphi)^{3/2}}$$

$$N = \frac{a}{(1-e^2\sin^2\varphi)^{1/2}}$$

Abridged Molodensky method: - The Abridged Molodensky method is a simplified version of the Molodensky method. The equations are

$$M\Delta\varphi = -\sin\varphi\cos\lambda\Delta X - \sin\varphi\sin\lambda\Delta Y + \cos\varphi\Delta Z + (a\Delta f + f\Delta a)\cdot 2\sin\varphi\cos\varphi$$

$$N\cos\varphi\Delta\lambda = -\sin\lambda\Delta X + \cos\lambda\Delta Y$$

$$\Delta h = \cos\varphi\cos\lambda\Delta X + \cos\varphi\sin\lambda\Delta Y + \sin\varphi\Delta Z + (a\Delta f + f\Delta a)\sin^2\varphi - \Delta a$$

Grid-based Transformation Method

Grid-based methods allow modeling the differences between the systems and are potentially the most accurate method. The area of interest is divided into cells. The National Geodetic Survey (NGS) publishes grids to convert between NAD 1927 and other older geographic coordinate systems and NAD 1983. Accuracies can vary depending on how good the geodetic data in the area was when the grids were computed.

Projection

Maps are flat, but the surfaces they represent are curved. Transforming three-dimensional space onto a two-dimensional map is called projection. Projection formulas are mathematical expressions that convert data from a geographical location— latitude and longitude—on a sphere or spheroid to a representative location on a flat surface.

Map projections are systematic transformations of the spheroidal shape of the Earth so that the curved, three-dimensional shape of a geographic area on the Earth can be represented in two dimensions, as x,y coordinates.

A projected coordinate system is defined as a flat, two-dimensional surface. Unlike a geographic coordinate system, a projected coordinate system has constant lengths, angles, and areas across the two dimensions. A projected coordinate system is always based on a geographic coordinate system that is based on a sphere or spheroid.

In a projected coordinate system, locations are identified by x,y coordinates on a grid, with the origin at the center of the grid. Each position has two values that reference it to that central location. One specifies its horizontal position and the other its vertical position. The two values are called the x-coordinate and y-coordinate. Using this notation, the coordinates at the origin are x = 0 and y = 0 (Fig. 6).

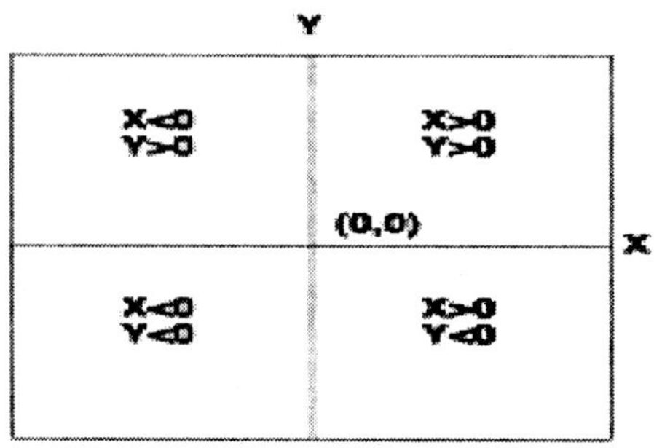

Fig.6 Projected Coordinate System

On a grided network of equally spaced horizontal and vertical lines, the horizontal line in the center is called the x-axis and the central vertical line is called the y-axis. Units are consistent and equally spaced across the full range of x and y. Horizontal lines above the origin and vertical lines to the right of the origin have positive values; those below or to the left have negative values. The four quadrants represent the four possible combinations of positive and negative x- and y-coordinates. When working with data in a geographic coordinate system, it is sometimes useful to equate the longitude values with the X axis and the latitude values with the Y axis.

Mapping and Map Layout

Cartography can be described as the graphic principles supporting the art, science, and techniques used in making maps or charts. It was developed in a time before the computer and GIS. Traditionally, maps have been created to serve two main functions. The first function has been to store information. Creating a map has been a way to record information for future reference. The second function has been to provide a picture to relay spatial information to a user.

The purpose for designing a map is critical to its design. When designing a map, a map maker needs to know the answers to some fundamental questions, such as: What is being mapped? Who is the audience? How is this map being presented, on its own or as part of a report? What medium will be used to display this map?

Map formats: Generally maps are considered to be in two formats. One is a general reference map such as a USGS topographic map or the map of a city. In this form the map is providing information to convey where things are in relation to each other. The second is a thematic map, where the map is used to convey information about a particular theme or multiple themes, such as land use, population, or health statistics.

Basic mapping principles: There are many kinds of maps, each with general and possibly specific requirements. While a skilled cartographer is usually required to make maps with specific or special requirements, anyone can make good general and informative maps by considering the following simple guidelines. These guidelines have been organized into seven areas that you can use as a checklist for creating or improving your maps.

- Purpose—Typically, a map does not have more than one purpose. Trying to communicate too much in one map—having more than one purpose for the map—tends to blur the message and confuse the map reader. Using two or more maps, each focused on a single message, is always a better strategy.
- Audience—who will be reading your map? Are you designing a map for a few readers or for a large audience of hundreds or millions of people? It's better to target your map to the person least prepared to understand your map's message.
- Size, scale, and media—The physical size of a map relative to the geographic extent shown on the map will dictate the scale of

the map and determine how you will represent the actual size and number of features shown on the map. Data is often collected at a particular scale. However, if you're not displaying the data at that scale, be sure your data "fits". For example, roads typically collected for 1:24,000 mapping will be far more detailed than needed for a smaller scale map (such as 1:2,000,000), so be sure to reduce the number of roads drawn on your map. Media also plays an import role, because a map printed on newspaper will not show fine details clearly, whereas one printed on high-quality paper will. In addition, the details on a digital map could vary depending on the viewing program. For example, a static map used in a Web page would be designed to encompass less information than one designed for browsing using a program such as ArcReader.

- Focus—Refers to where the designer wishes the map reader to first focus. Typically cool colors (blues, greens, and light gray) are used for background information, and warm colors (red, yellow, black) are used to capture the reader's attention.
- Integrity—you may want to cross-validate some of your information, such as the names or spelling of some features. If the data was produced by another organization, it is customary to give that organization credit on the map.
- Balance—how does your map look on the page or screen? Are the parts of the map properly aligned? The body of the map should be the dominant element. Try to avoid large open spaces. Be flexible in where you place elements (that is, not all titles need to go at the top). Should some components on the map be contained within a border?
- Completeness—a map generally should contain some basic elements, such as a title, legend, scale bar, and North arrow; however, there are exceptions. For example, if a graticule exists, it is not necessary to place a North arrow. Basically, place all the information you think your readers need to fully understand the map. Examples of the basic elements to include on a map are shown in the diagram titled 'Map elements diagram'.

Before publishing your map, it is always a good idea to have someone else look it over, especially for spelling and overall appearance.

Map Element

- Map body—the primary mapped area. You can display more than one image of your primary mapped area within your document. For example, you may want to portray change by showing several images with differing but related information, such as population maps of various years. Your map may also contain a locator map (a smaller-scale map used to help the reader in understanding where the main area of interest is located), an inset map (used to give more detailed information of an area within the main map that may not easily be understood), or an index map (often used to show where in a series of maps one map exists). All are used to assist in communicating your information to others. In ArcMap, each of these mapped areas is referred to as a data frame.
- Title—is used to tell the reader what the map represents. This is often placed on a map layout as text.
- Legend—lists the symbology used within the map and what it represents. This can be created using the Legend Wizard in the layout and edited further once created.
- Scale—Provides readers with the information they need to determine distance. A map scale is a ratio, where one unit on the map represents some multiple of that value in the real world. It can be numeric (1:10,000), graphic (a scale bar), or verbal (one inch equals 10,000 inches). Maps are often referred to as "large" or "small" scale. This "size" reference refers to the ratio (or fraction). For example, a 1:100 scale map is larger than a 1:10,000 map, because 1/100 (0.01) is a larger value than 1/10,000 (0.00001). A smaller scale map displays a larger area, but with less detail. The scale is inserted in the map layout view.
- Projection—is a mathematical formula that transforms feature locations from the Earth's curved surface to a map's flat surface. Projections can cause distortions in distance, area, shape, and direction; no projection can avoid some distortion. Therefore, the projection type is often placed on the map to help readers determine the accuracy of the measurement information they get from the map.
- Direction—this is shown using a North arrow. A map may show true north and magnetic north. This element is inserted in the map layout view.

- Data source – the bibliographic information for the data used to develop the map.

Other map components include (but are not limited to) dates pictures, graticules or grids, reports, tables, additional text, neat lines, and authorship.

REFERENCES

Aerospace, (1968). The 1968 Aerospace Year Book. Washington, D.C.: Books, Inc.

Agrawal, N. K., (2004). Essentials of GPS. Hydrabad: Spatial Networks Pvt. Ltd.

Buckeye, K.R., (2004). GIS - GPS for Distance - Based User Fee System, GIS Development, 8 (11), pp. 34-37.

Burchell, Wade, Lt. Cmdr, (1997). Reserve Helos Get Magic Lantern Naval Institute Proceedings, 74-75

Curtis Peebles, (1987). Guardians: Strategic Reconnaissance Satellites, Novato, CA: Presidio.

Galileo (2005). Galileo Home Page at http:\\Galileo-pgm.org

Ganesh, A. and Narayanakumar R., (2003). Principles of GPS. Tiruchirappalli: Bharathidasan University.

Gartner, George, V. Radoczky and G. Retscher, (2005). Location Technologies for Pedestrian Navigation, GIS Development, 9 (4), pp. 22-31.

GLONASS, (2005) GLONASS Home Page at http:\\rssi.ru/SFCSIC/SFCSIC_main.html.

Gouzhva, Yuri G. and Arvid G. Guevorkyan, Arkady B. Bassevich, Pyotr P. Bogdanov, (1992). High-Precision Time and Frequency Dissemination With Glonass, GPS World, 3, (7), pp. 40-49

GPS, (2005). GPS Home Page at US Coast Guard www.navcen.useg.gov./default.htm

GPS Word, (1992). Protesters invade Rockwell Plant, Damage GPS Satellite, GPS World, 3 (6), p. 12

GPS Word, (1992). Satellite Set Healthy, Launch Planned for Mid-June, GPS World, 3 (6), p.12

GPS Word, (1993). FAA Approves GPS for IFR Operations, GPS World, 4, (7), p.18

He, Xiufeng, W. Sang, Y. Chen and X. Ding (2005) Steep – Slope Monitoring. GPS Multiple Antenna System at Xiaowan Dam, GPS World, 16 (11), pp. 20-25.

Higbie, Paul R. and Norman K. Blocker. (1994). Detecting Nuclear Detonations with GPS, GPS World, 5 (2), pp.48-50

Hofmann-Wellenhof, B. Lichtenegger, H and Collins.J, (2001). GPS Theory and Practice, New York: Springer Wien.

Ivan A. Getting. (1993). The Global Positioning System, IEEE Spectrum, pp.36-47

Jethwani, Hiresh, (2004). GPS Navigation in Indian Automobiles, GIS Development, 8 (11) p. 44.

Johnson, Nicolas, L., (1994). GLONASS Spacecraft, GPS World, 5 (11), pp. 51-58

Leica (2003). GPS Basics: Introduction to GPS, Leica Version 1.0.

Leick Alfred, (2004). GPS Satellite Surveying, Third Edition, New Jersey: John Wiley and Sons Inc.

Kaplan E.D. (Ed). (1996). Undernawy GPS Principles an Application. Norwood. MA: Artech House.

Kennedy, Michael (2002). The Global Positioning System and GIS : An Introduction, London : Taylor and Francis

Kraak, Menno-Jan and Allan Brown, (2001). Web Cartography. London : Taylor and Francis.

Parkinson, Bradford, W., (1994). GPS Eyewitness: The Early Years. GPS World, 5 (9), pp. 32-45.

Schiestl, Andrew J., (2004). GPS and Alarka Rail Safety. GPS World, 15 (3), pp. 14-19.

Stine, G. Harry (1971). ICBM. New York: Orion Books.

Syed, Salmand and E. Cannon (2005). Map-Added GPS Navigation, GPS World, 16 (11), pp. 39-44.

Tsui, James Bao-Yen, (2005). Fundamentals of Global Positioning System Receivers. A Software Approach, Texas: Wiley.

Whitford, Marty, (2005) Hurricane Hunters : GPS Dropsondes Trace Katrina's Course, GPS World, 16 (11), pp. 20-22.

Worthman, Ernest, (1997). Global Positioning Systems - Technology into the 21st Century, RF Design, pp. 44-54

Wysocki, Joseph, Lt. Col., (1991). GPS and Selective Availability - The Military Perspective. GPS World, 2, (7), pp. 38-44

Xie, Yicheen, D. Goby, R. Raymond, R. Pizzi, (2004). Going Beyard Automatic Vehicle Location, GIS Development, 8 (11), pp. 30-33.

Web Resources

http://www.aero.org/education/index.html

http://www.buzzle.com/editorials/9-10-2004-115.asp?viewPage=2

http://www.colorado.edu/geography/gcraft/contents.html

http://www.colorado.edu/engineering/ASEN/asen5090/GPS3.sm.jpg

http://igscb.jpl.nasa.gov/overview.html

http://www.gmat.unsw.edu.au/snap/gps/gps_survey/chap2/211.htm

http://www.gmat.unsw.edu.au/snap/gps/gps_survey/chap3/311.htm

http://www.gps-practice-and-fun.com/personal-location-tracking.html

http://hyperphysics.phy-astr.gsu.edu/hbase/gpsrec.html

http://www.montana.edu/places/gps/understd.html

http://www.navcen.uscg.mil/loran/loranff.htm

http://www.navcen.uscg.mil/omega/omegaff.htm

http://www.navcen.uscg.mil/gps/gpsff.htm

http://www.palowireless.com

http://science.howstuffworks.com/nuclear2.htm

http://www.trimble.com/gps/how.html

APPLICATIONS OF GEOSPATIAL TECHNOLOGY
A. Ganesh (Ed.), (2006) : pp. 267-290
Satish Serial Publishing House, New Delhi
E-mail: ssp@ndf.vsnl.net.in, hkjain1975@yahoo.com
website: www.satishserial.co

20

Distributed GIS–Static Map Publishing for Bharathidasan University

K. SUBADEVI AND A. GANESH

The development of GIS has been influenced by the progress of IT. The Internet influenced GIS in three major areas, as data access, spatial information dissemination and GIS modeling/processing. There are three kinds of internet GIS and has many different faces and alternative names such as 'Internet Mapping', 'Distributed GIS', 'Web-based GIS', 'Networked-GIS', 'On-line GIS' etc., Variety of Internet-GIS is used for data sharing, information sharing and knowledge sharing (Ming-Hsiang Tsou, 2004). Among them information sharing is an application for 'Web–Based Map Display'. Static map publishing is a simple and effective way to communicate the user, for thin clients, that preliminary stage of evolution of distributed GIS. It supports the solution for decision-making in all fields through spatial or non-spatial query. Thus, Distributed GIS is a public resource of general information. GIS can be regarded as the high tech equivalent of the map. The main aim of the present study is to publish static map of Bharathidasan University for students as well as administrators and to disseminate distributed GIS.

Methodology

Exclusive GPS Survey for entire University Campus, Assigning attributes for all point and polyline features captured through GPS survey, Integrating GIS/GPS for effective mapping, Integrating all these information with respective high-resolution satellite image to pertain synoptic view of the university area. Finally, based on this information a query platform is constructed to create different queries by administrators as well as others using ArcGIS 9 Version with evolving a Distributed GIS for University and create URL for map publishing and to publish static maps to access the users.

Data Base

GPS Survey records using single frequency Leica GS5, Satellite Images: IRS- 1D LISS III PAN merged scene includes 58 J / 9,10,13,14 toposheet area and Collateral information.

Software and Instruments Used

ERDAS IMAGINE 8.5 – To mosaic the satellite images and create AOI (Area Of Interest) file, ArcGIS 9 version – For editing, database creation, query and mapping, Micro active sync – To integrate the GIS / GPS and synchronize the data and Leica GS5 – GPS used to survey along with the Pocket PC.

Study Area: Bharathidasan University

Location and Extent

The study area for the present study is the Bharathidasan University's Main Campus, which lies between 10° 39′ 44″ and 10° 41′ 12″ North latitudes and 78° 44′ 16″ and 78° 45′ 2″ East longitudes. Bharathidasan University is located in the Tiruchirappalli district and situated on the Tiruchirappalli-Pudukkottai Highways (NH210), which is 16 km north of Tiruchirappalli Junction in the Suriyur Block of Tiruchirappalli Taluk (Fig. 1). The total area is 631.17 acre.

Administrative Division: The present study area comes under the Suriyur block of the Tiruchirappalli District.

Physiography: The details relating to the physiography of the study area is obtained from the Survey of India topographic sheets of 1:50,000 scale. The study area in general is a gently undulating towards south. The slope gradually decreasing towards the southern part from 100 to 80 m contour interval.

Geomorphology: The geomorphic feature of the study area is interpreted from IRS-1C LISS III satellite data. Based on the image interpretation elements, the three geomorphic features such as colluvial fills in north eastern boundary until near percolation pond adjacent to Helipad, Pediments found in the south east centre of University boundary and pediment on gneiss are found to occur in the rest of campus.

Geology: The study area is composed of Hornblende Biotite Gneiss. It belongs to the Achaean range and covered by magnetite.

Soil: The soil map of the study area was generated by making use of the relevant details from the report published (1998) by the Soil Survey and Land Use Organization of the Department of Agriculture, Government of Tamil Nadu. It composed of the Paralithic Haplustalf (Phsf) class (Manjula and Aruchamy, 2004).

Bharathidasan University

The Government of Tamil Nadu established Bharathidasan University, named after the revolutionary Tamil Poet, Bharathidasan, in the year 1982 through the Bharathidasan University Act, 1981. It is situated at the outskirts of Tiruchirappalli, a city of rich cultural heritage and has cluster and cluster of reputed educational institutions. The University is a member of the Association of Indian Universities and Association of Commonwealth Universities. It is entitled to receive grants and donations from University Grants Commission (UGC) and other agencies all over the world.

The University undertakes teaching, research and extension services through its Departments and Schools of Excellence and 81 affiliated colleges spread over seven districts of Tamil Nadu, viz., Tiruchirappalli, Karur, Thanjavur, Nagapattinam, Tiruvarur and Pudukkottai.

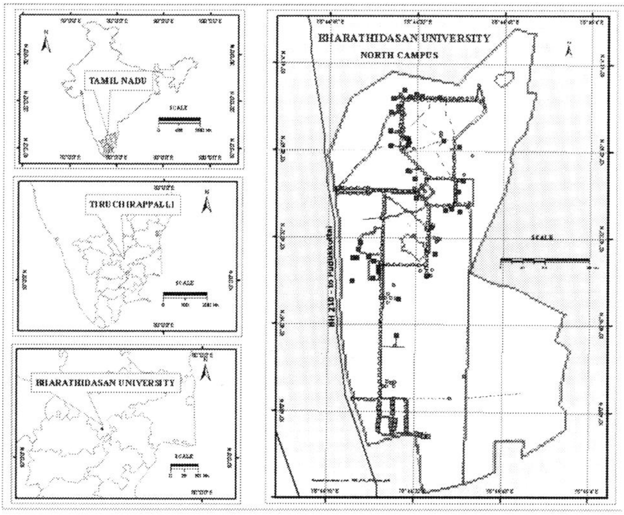

Fig.1 Bharathidasan University Location

Main Campus

The Main Campus of the University is at Palkalaiperur, which is situated at about 16 km. from Tiruchirappalli on the Tiruchirappalli - Pudukkottai National Highway 210 (Fig. 2-6). The departmental buildings are sited in a tract of land with an extent of 631.17 acres. The Departments in the Main Campus are: Mathematics, Physics, Chemistry, Computer Science, Geology, Geography, Commerce, School of Life Science, Animal Biotechnology, Eco-biotechnology, Plant Science, English, Tamil, Economics, Sociology, Social Work, National Facility for Marine Cyanobacteria, Centre for Nonlinear Dynamics, History, Computer Centre, Informatics Centre, Central Workshop, Yoga Centre and Physical Education. Hostels for Men and Women, Library, University Health Centre, State Bank of India Post Office, Telecom Centre and Telephone Exchange are also located in this Campus.

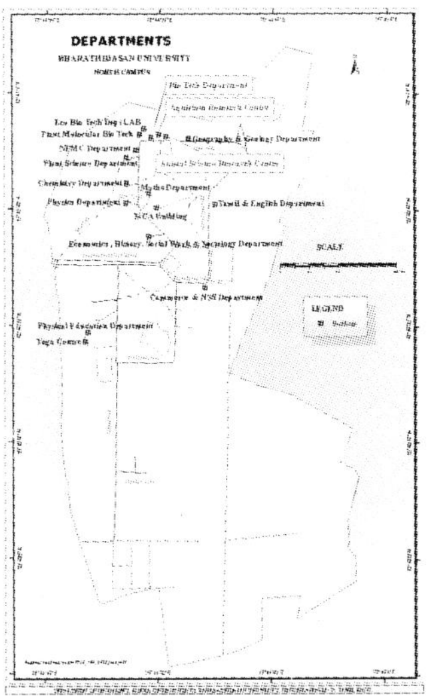

Fig.2 Departments

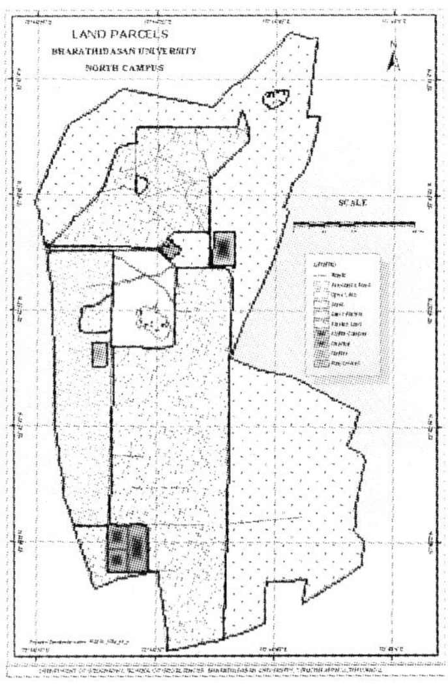

Fig.3 Land Parcles

Khajamalai Campus

The University has a down town campus known as Khajamalai Campus with an extent of 52 acre. The Academic Staff College, Community College, School of Energy, Centre for Remote Sensing, Department of Educational Technology, Centre for Distance Education, Center for Adult Education and Extension Counter of SBI, institute of Career and Entrepreneur Development, Technology Park and a hostel for men are located. Bharathidasan Institute of Management (BIM) established by the University, is functioning at BHEL premises at Thiruverumbur, Tiruchirappalli.

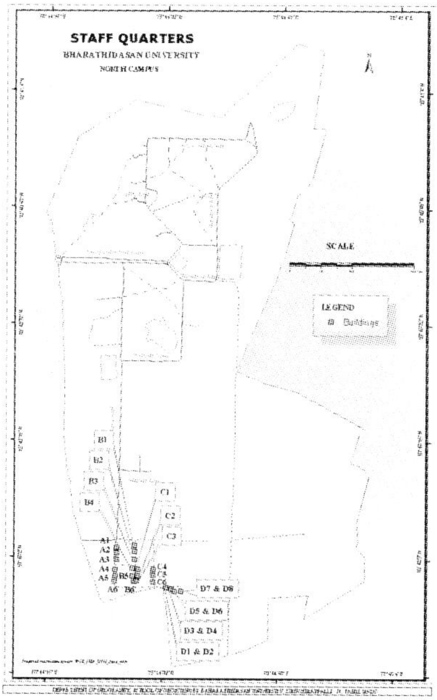

Fig.4 Staff Quarters

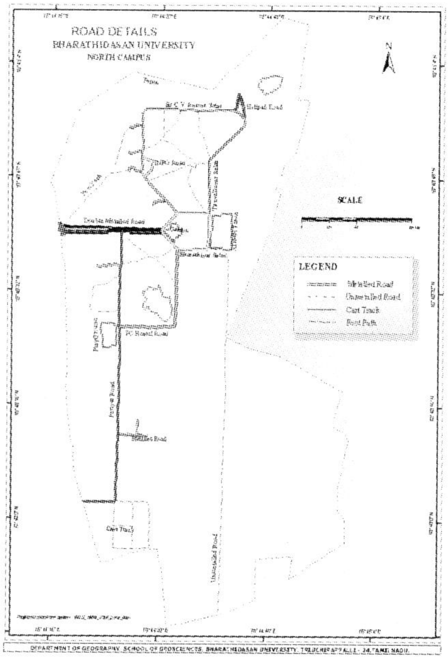

Fig.5 Roads

School of Engineering and Technology (SET) Campus

The School of Engineering and Technology of Bharathidasan University is located in 500 acres of land on the south of Main Campus, was established in 1999 with a view to impart high-tech programmes including Information technology, petrochemical technology, biotechnology and pharmaceutical engineering and technology.

GPS SURVEY

The Global Positioning System (GPS) is a satellite - based radio-navigation system established by the U.S. Department of Defense for military positioning applications and as a by-product, has been made available to the civilian community. GPS is a complex system, which can be used to achieve position accuracies ranging from 100 m to a few millimetres depending on the equipment used, and procedures followed. In general, higher accuracies correspond with higher costs and more complex observation and processing procedures. GPS receiver manufacturers to come up with low cost solutions for GPS surveying.

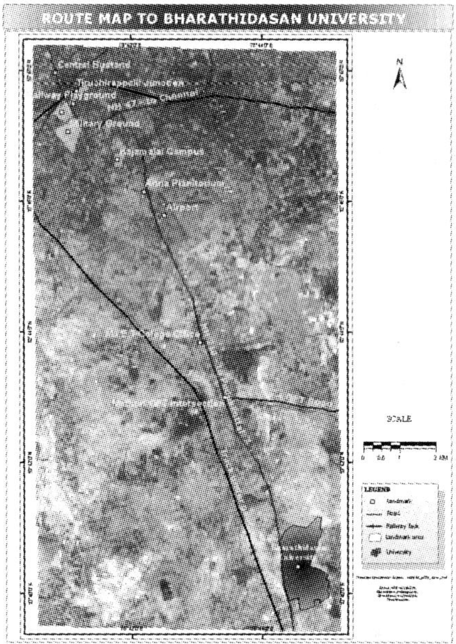

Fig.6 Route to the University on IRS IC Image

GPS Used for Survey

The GPS used in the study is a single frequency Leica GS5 receiver whose accuracy is 30 m in stand-alone mode connected with mobile computing device, Compaq iPAQ™ Pocket PC running Windows CE platform through ArcPad version 6.0, mobile GIS software and it is connected to mobile GPS with external antenna. The ArcPAD from ESRI is mobile GIS software work with most popular GPS receivers that support the data transfer through NMEA protocol version 2.0.

Pre-Survey Plan/Preparation

GPS survey requires different planning, execution and processing techniques. Project planning is one of the most important aspects of GPS surveying, as careful planning maximizes the chances of the survey achieving the desired accuracy, within a reasonable time and to budget. The nature and aim of the survey considerations are discussed, which includes do and don't procedures while surveying. Shape files were created to capture the point and Polyline features and assign WGS 84_UTM_Zone_44N projection.

Shapefiles then exported into ArcPad in Pocket PC through creating partnership and synchronize the data using Micro Active sync software. For record attributes of object the field data record sheet had prepared that consists GPS-ID, Feature type, PDOP and Remarks. In addition to this, Operator, Assistant, Recorder, instrument, Date and layer name are included. Field visit made to organize the survey plan to avoid time consumption. Since the total area is nearly 1000 acres have to be surveyed, vehicle needs to capture the Polyline like road. Therefore, a vehicle is arranged for survey and other instrument, communication kits.

GPS Survey

The survey has been started with creating shape files of appropriate feature class i.e. point and line with assigning unique identification number for the each feature. The method of survey can classify as single point, kinematic real time survey method. Coordinates of linear features and point features are captured either using vehicle (Receiver is mounted with vehicle) or by walking with receiver wherever the vehicle is not possible such as fence, playground, pond etc. The base map of the university area has been loaded with Tiruchirappalli district map at the background for the rationale of planning the survey.

Satellite geometry has a direct effect on positioning accuracies, which is quantified by the geometrical dilution of precision (GDOP). The best positioning accuracies are achieved when satellites have good spatial distribution in the sky. For GPS positioning, the lower the PDOP is better with an increased number of satellites and vice versa. A PDOP below 5 or 6 is generally the recommended upper limit for positioning, particularly for short occupation times (e.g. a few minutes). To maintain the accuracy of the survey the PDOP have maintained within 4 as a average throughout the survey.

GPS receivers, processing software, or both may have an option to set a specific mask angle also referred to as cutoff angle. The mask angle refers to the elevation angle below which GPS signals will not be recorded. In present survey the mask angel have maintained as 15 degree. A satellite is said to be visible if it is above the specified mask angle for the time and location of interest assuming no obstructions are present.

Totally the survey took 15 hours to cover nearly 1000 acres of land, radically spotted with numerous buildings for various departments, administration, library, hostel, post office, bank etc.,

Besides buildup features, network of street light, culverts, sumps, foot path network between various buildings also surveyed. A field data record sheet has used to document the PDOP value and remarks if any during the survey and it is used during the post processing of feature conformation. Totally 336 points are captured and maintained with unique object ID that starts from 2001 and followed. Similarly 90 line features captured and ID gave from 1001, which are all recorded in field data record sheet with remarks.

Integration of GPS and GIS

Combining the GPS data with GIS allows for greater capabilities than what GPS and GIS can provide individually. With the combination of two technologies one is able to display the 'FIELD/ ACTUAL SITE' on a PC and make informed decisions. There is no need to make specific site visits or review several documents/drawings. Also, another benefit of the integration is the fact that the data can be shared by unlimited users in various departments for their own specific needs and analysis.

Another important advance in this technology has been the introduction of software, which allows bringing into GIS not only GPS position information but a digital picture i.e. satellite image of the same area. With the software, one can study relationships between features and also view actual photographs of the features on the PC.

Process of Integration

Once GPS readings are available they need to be processed further before being converted into desired format such as maps. Both GPS and GIS are system executed through software, all the processes of integration have to flow through them. There are a variety of GIS software available in the market, viz., ArcMap, Arc View, Geomedia, etc. Some GPS instruments like the products from Leica are executed through the software ArcPad. The processes of integration thus involve the modules of both GPS and GIS software.

The following are the major processes involved in the integration.

1. Projection
2. Editing and Validation
3. Attribute Data Management
4. Query - spatial or Non - spatial
5. Mapping and Map Layout

Projection

Maps are flat, but the surfaces they represent are curved. Transforming three-dimensional space onto a two-dimensional map is called projection. Projection formulas are mathematical expressions that convert data from a geographical location— latitude and longitude—on a sphere or spheroid to a representative location on a flat surface. Map projections are systematic transformations of the spherical shape of the Earth so that the curved, three-dimensional shape of a geographic area on the earth can be represented in two dimensions, as x, y coordinates.

A projected coordinate system is defined as a flat, two-dimensional surface. Unlike a geographic coordinates system, a projected coordinates system has constant lengths, angles and areas across the two dimensions. A projected coordinate system is always based on a geographic coordinate system that is based on a sphere of spheroid. A sphere is based on a circle, while a spheroid (or ellipsoid) is based on an ellipse. An earth-centered, or geocentric, datum uses the earth's center of mass as the origin. The most recently developed and widely used datum is WGS 1984. It serves as the framework for locational measurement worldwide.

Universal Transverse Mercator (UTM) system is a specialized application of the Transverse Mercator projection. The globe is divided into 60 north and south zones, each spanning 6° of longitude. Each zone has its own central meridian. The present study area extent lies between 10° 39′ 44″ and 10° 41′ 12″ North latitudes and 78° 44′ 16″ and 78° 45′ 2″ East longitudes, WGS 1984 UTM 44 zone has been used as projection to the Shapefiles while prepared for survey and loaded with the Tiruchirappalli district map at the background for rationale purpose.

Editing and Validation

There are generally two different types of errors could be anticipated from a GPS reading, viz., errors related to the position of the point, errors related to linear feature. Point positioning errors may be due to varied reasons. These errors related to points could be solved by precise positioning techniques, observing the same point at different time periods and satellite availability. Errors related to linear features are of two types, viz., Overshoot and Undershoot. Overshoot is the line gone beyond its target snap position. While Undershoot is related to the line which is felt short of some distance before its target snap position. Overshoot could be corrected by trimming the extended line while the undershoot could be snapped to the target snap position by extending the line segment.

In present study, GPS survey started with errors except point feature captures. The line feature captured mostly along with the overshoot with several bit of line segments due to the twisted of the receiver, which can found prominently for road networks, particularly the unmetalled road, pond and fence. Similarly, Undershoots are identified with playground, administration building, garden etc., Errors are edited with aid of Editor tool in ArcGIS. Some vertices could add and deleted in accordance with requirements. The Advanced Editor tool is manifesting editing task. Merge option is effective to group the segment of line features that has to be considered common entity like unmetalled road, metal road, fence and so on.

Data Management

In GIS, there are two types of data to be managed: spatial data and attribute data. An entity (point, line or area) has both spatial and attribute data to describe it. Spatial data are the 'where things are' data and attribute data the 'what things are'. For example, a latitude and longitude reference gives the location of a point entity and to accompany this there would be attribute data about the nature of the real-world feature the point represents. Conventionally, databases have stored only non-spatial entities; however, all entities in GIS are spatial. They can help to convert data into information with value. So the database is a set of structured data.

Data in a computer database are managed and accessed through a database management system (DBMS). There are a number of different database data models. At present the relational database model dominates GIS. It uses to handle spatial as well as attribute data and based on concepts proposed by Codd (1970). Queries are possible on individual tables or on groups of tables. There are few restrictions on the types of queries possible. SQL (Structured Query Language) has been developed to facilitate the querying of relational database. However, SQL has not really developed to handle geographical concepts such as 'near to', 'far from' or 'connected to'. Relational databases are predominantly used for the handling of attribute data in GIS (Heywood, 2001).

For the present study, attributes have been collected during the survey in the field record sheet to the spatial object. ArcGIS extensions using intelligent RDBMS concept to capture, store and retrieve the data in database through maintain the object ID, geometry of the entity with updating. In addition to this, collect other primary data related to the spatial features from Estate Maintenance and information regarding departments and others collected through the questionnaire from corresponding offices. The collateral information had also created in MS Excel and related to the attributes for query purpose. The unique ID has given to each object in the data capturing. Whereas the common unique ID has been provided to ease retrieval of the common features such as 10 for Departments and Centers, 20 for amenity and so on in case of point features. Similarly, line features are classified as six and provide ID range from 1 to 6. For symbolize the features sake, features are selected on the basis of SQL query builder using basic logical operator expressions and exported as new shape files.

Query - spatial or Non – spatial

Query is an input to the computer that aims to retrieve data from a database. This is done with the aid of a database management system (DBMS) (Lo, 2002). A special nature of data exploration in GIS is the linkage between spatial data and attribute data. The linkage between spatial and attribute data is an important characteristic of the vector data model. Using the feature attribute table, a vector – based map can be joined or linked database for data query.

Spatial data query refers to the process of retrieving data from a map by working with map features. It is the geographic interface to the database and is therefore useful for tasks that cannot be easily accomplished by attribute data query. As in attribute data query, results of spatial data query may be visually inspected or saved as new maps for further processing.

Attribute data query retrieves a data subset from a map by working with its attribute data. The selected data subset may be visually inspected or saved for further processing. Attribute data query requires the use or expressions, which must be interpretable by a GIS of a database management system (Chang, 2002). SQL (Structured Query Language) is a standard computer language for accessing and managing databases. It is a powerful language; used to define one or more criteria that can consist of attributes, operators, and calculations and allows selecting features on a given layer using a SQL expression (ArcGIS 9 Manual, 2004).

In the present study, spatial query used to extract the features to classify under the premise viz., built-up features grouped into amenity, departments, quarters and others. In order to make analysis the data, particularly adjacency modeling functions like allocating, suitability analysis this would be prime step. Data queries in ArcGIS follow Boolean Algebra (after the English logician George Boole, 1815-1864) and consist of logical expressions and Boolean connectors, which are facilitated feature selection and endow with result of query (Fig. 7-9).

Mapping and Map Layout

Spatial features are characterized by their locations and attributes. To display a spatial feature on a map, use a map symbol to indicate the feature's location and a visual variable, or visual variables, with the symbol to show the features attribute data. Choosing the appropriate map symbol and visual variables is the main concern for data display. Visual variable in cartographic symbolization include shape, size, texture, pattern, hue, value and chroma. (Chang, 2002).

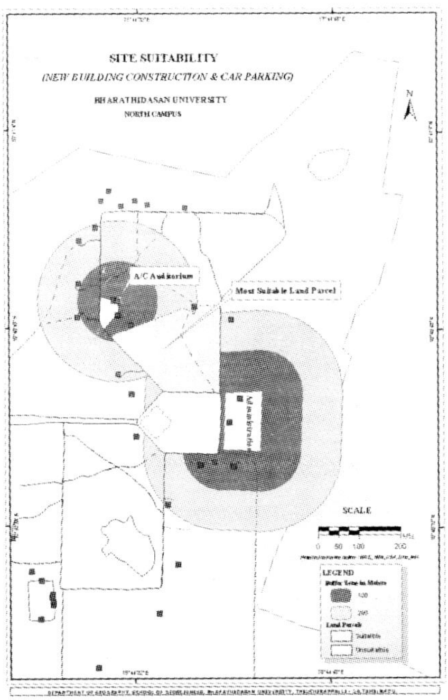

Fig.7 Site Suitablity

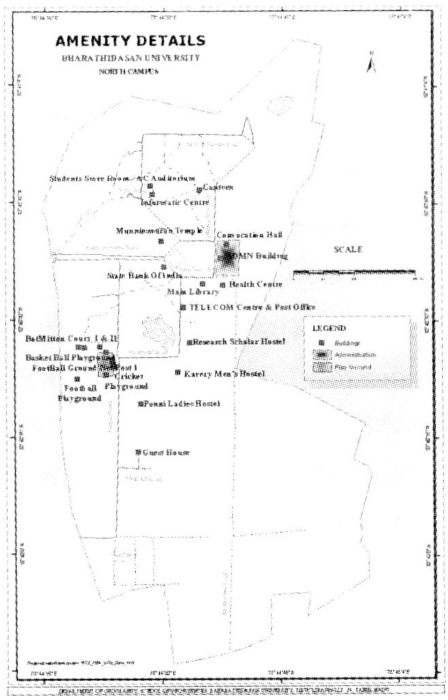

Fig.8 Amenities

Cartography can be described as the graphic principles supporting the art, science, and techniques used in making maps or charts. It was developed in a time before the computer and GIS. The purpose for designing a map is critical to its design. ArcGIS offers choices of all symbolization elements.

When designing a map, a mapmaker needs to know the answers to some fundamental questions, such as: What is being mapped? Who is the audience? How is this map being presented, on its own or as part of a report? What medium will be used to display this map? These questions are observed for mapping to the present study.

Basic Mapping Principles: There are many kinds of maps, each with general and possibly specific requirements. In order to make maps with specific or special requirements, the following simple guidelines should considered. These guidelines have been organized into seven areas that are creating or improving maps.

Purpose: Typically, a map does not have more than one purpose. Trying to communicate too much in one map—having more than one purpose for the map—tends to blur the message and confuse the map-reader. Using two or more maps, each focused on a single message, is always a better strategy. Except the base map with all features that consists 336 points and all line features, all other are used to convey the single theme on the basis of student and administrative purpose.

Audience: Who will be reading the map? Are you designing a map for a few readers or for a large audience of hundreds or millions of people? It's better to target of map to the person least prepared to understand your map's message. Since the present work's output map is to be expected to use by students and administration purpose the simple style has followed to design.

Size, Scale, and Media: The physical size of a map relative to the geographic extent shown on the map will dictate the scale of the map and determine how it will represent the actual size and number of features shown on the map. Data is often collected at a particular scale. Media also plays an import rôle, because a map printed on newspaper will not show fine details clearly, whereas one printed on high-quality paper will. In addition, the details on a digital map could vary depending on the viewing program. The output has prepared to publish as static map in BARD (Bharathidasan University) personal site for users. Though used in web page, with aid of ArcGIS advantages maintain these elements.

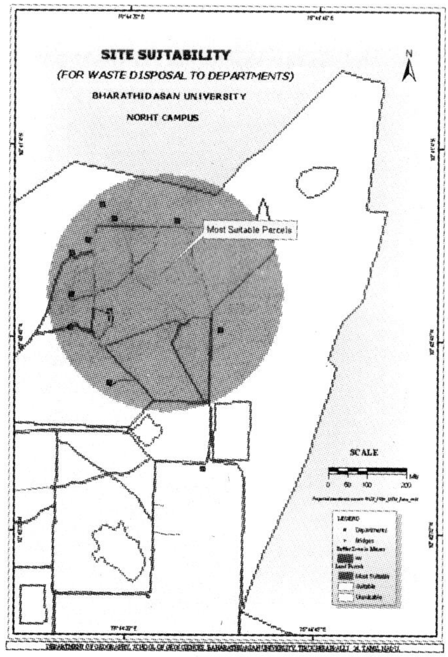

Fig.9 Site Suitablity for Waste Disposal

Focus: Refers to where the designer wishes the map-reader to first focus. Typically cool colors (blues, greens, and light gray) are used for background information, and warm colors (red, yellow, black) are used to capture the reader's attention. It is taken into account to focus the theme, for instance yellow used to background with gradient shade, all labels are depicted by black colour and so on.

Integrity: It is need to cross-validate some information, such as the names or spelling of some features. If the data was produced by another organization, it is customary to give that organization credit on the map. Since all spatial and it's related data has been collected as first hand, even survey and collateral data, no necessary to add. But the satellite imagery used in Route map to Bharathidasan University, which is provided by Department of Geography that has integrities.

Balance: How does your map look on the page or screen? Are the parts of the map properly aligned? The body of the map should be the dominant element. Try to avoid large open spaces. Be flexible in where you place elements that is, not all titles need to go at the top. Should some components on the map be contained within a border? Because of the study area in elongated rectangle in shape the placement of the scale bar and legend used to fill the space. In addition to this, annotation in map features used to keep the label as legible. Annotation is textual and graphical information that either labels the map and features presented within it, or provides the user with supplementary information (Heywood, 2001).

Completeness: A map generally should contain some basic elements, such as a title, legend, scale bar, and North arrow; however, there are exceptions.

Map Layout

Map design is a visual plan to achieve a goal or the purpose of map. The purpose of map design is to enhance map communication, which is particularly important for thematic maps. A well - designed is balanced, coherent, ordered and interesting to look at. Layout is the arrangement and composition of map elements on a map such as the map body, title, legend, scale and north arrow.

Map Elements

Map Body: This is the primary mapped area. You can display more than one image of your primary mapped area within your document. For example, to portray change by showing several maps that contains a locator map, a smaller-scale map used to help the reader in understanding where the main area of interest is located. All are used to assist in communicating information to others. In Arc Map, each of these mapped areas is referred to as a data frame.

Title: Title is used to tell the reader what the map represents. This is often placed on a map layout as text. Similar style of text has followed for keeping appreciation.

Legend: Legend lists the symbology used within the map and what it represents. This can be created using the Legend Wizard in the layout and edited further once created. It offers to determine the items to be placed in the legend

Scale: It provides readers with the information they need to determine distance. A map scale is a ratio, where one unit on the map represents some multiple of that value in the real world. It can be numeric (1:10,000), graphic (a scale bar), or verbal (one inch equals 10,000 inches). Maps are often referred to as 'large' or 'small' scale. A larger scale map displays a small area, but more details. The present study area map scale is large scale map and covers large details as well as graphic scale has inserted for easy understanding, measure if necessary.

Direction: This is shown using a North arrow. A map may show true north and magnetic north. This element is inserted in the map layout view.

Projection: Projection is a mathematical formula that transforms feature locations from the earth's curved surface to a map's flat surface. Projections can cause distortions in distance, area, shape, and direction; no projection can avoid some distortion. Therefore, the projection type is often placed on the map to help readers determine the accuracy of the measurement information they get from the map. It is indicated left bottom corner of the all maps

Data source: The bibliographic information for the data used to develop the map. Only satellite image used map i.e., Route map to Bharathidasan University, source has been mentioned.

Other map components include but are not limited to dates, pictures, graticules or grids, reports, tables, additional text, neatlines, and authorship.

Distributed GIS

Distributed GIS evolved from centralized mainframe GISystems to personal desktop GIS to distributed GIServices. The mainframe GIS and the desktop GIS is also traditionally called GIS systems, while the distributed GIS is referred to as distributed GIServices. It represents a dramatic departure from the traditional two-tier client/server model and a broader framework including both internet GIS and mobile GIS. This term emphasizes the software characteristics of Internet GIS and mobile GIS, which are distributed and dynamic (Peng and Tsou, 2003).

Ahmed Shaig (2001) categories Web based GIS based on its source. Two types of sources could be identified: Single source GIS, i.e. data and operations from single website/server and Multiple Heterogeneous Sources i.e. data or operations from multiple sites/servers. Amongst the single source systems, static and dynamic pages could be further sub-classified. Static pages, limited in its ability to perform spatial operations, form a fair portion of Web Based GIS. Static pages are used here as referring to pages that sends predefined requests to a web server where the data has already been prepared. A good example would be the use of GIF images to display maps based on user clicks

(Canterbury, 1998). The main function of a web based GIS is visualisation of spatial data based on user queries (Peng, 1997). There are different methods used achieve this in each of these type of web pages.

Peng and Tsou (2003) suggest a classification of Web mapping technology into four classes: (1) static map publishing, (2) static Web mapping, (3) interactive Web mapping, and (4) distributed geographic information services (Claus Rinner, 2003). This is known as the 'Evolution of Distributed GIS'.

Evolution of Distributed GIS

Static map publishing distributes maps on the web page as static map images in graphic formats like Portable Document Format (PDF), GIF, or JPEG. It relied on the early stage of Web technology-a giant URL-based HTTP server-to hyperlink ready-made maps on the Web. Maps are usually part of the HTML document to enrich the contents of the document. Users cannot interact with the maps or change their display format in any way.

Static Web Mapping: It involves the use of HTML forms and the CGI to link the user input on the web browser with GIS or mapping programs on the servers. The GIS or mapping engines create the map based on the user's request and generate an image map on the fly. The new image is sent via HTTP back to the user on the Web browser. However, the HTTP form is text based and allows limited user input. Users cannot define or draw a circle or a square on the image maps. Several variations of CGI are developed to improve the performance of CGI, such as Netscape's NSAPI, Microsoft's ISAPI and ASP, NeXT/Apple's WebObjects, Javasoft's servlets, and fast CGI. But the interaction between the user and the maps on the Web browser is still limited.

Interact Web Mapping: The interactive web mapping is more interactivity and intelligence are added to the web client side by using scripts like dynamic HTML and/or client-side applications like plug-ins, ActiveX controls, and Java applets. Some user queries can be processed on the client side without sending requests to the servers. But this approach still requires HTTP connections and the web servers to mediate between software objects running on the client-side machines and the servers that store these objects.

Distributed GIServices: The last stage is the distributed GIServices where GIS components on the Web client side can directly communicate with other GIS components on the server without going through an HTTP server and CGI-related middle-ware. Distributed GIServices rely on the communication between CORBA/Java ORB or Microsoft's SOAP on the client side and the CORBA/IIOP and server-side Java or .NET/COM + technology in the Microsoft world

Static Map Publishing

Static Pages

Web based GIS utilising static web pages use predefined set of information that has already been published into a web page. These pages with the required elements mostly graphics and text are displayed on request of the user who follows the links from the home page (Geodata Resources, 2000). Besides it is named as cartographic visualization since it was simple and efficient for most of the common GIS operations. These pages are hence used mainly for display only purposes. Spatial data can be stored in the form of either raster or more recently vector data. Raster data are the more commonly used type for such sites through the use of compressed image formats such as GIF (Graphics Interchange Format) or JPEG (Joint Photographers Expert Group). Raster images can be used to display images through text hyperlinks or for a more visual linkage through image maps. Image map is a method used to embed hyperlinks into a portion of image creating a more visual interaction.

Static pages can be used to answer queries submitted by a user. However, such queries cannot be adhoc and information relating to the query must be already published into a web page. Hence if adhoc querying is required, that GIS should be implemented using dynamic WebPages. An example of web based GIS utilising static pages include, the Canterbury Tour Map (Canterbury, 1998), where it uses an underlying map in GIF format to provide links to images via an image map. Hence static pages are used in a web based GIS if the system is relatively simple, display only and do not require adhoc querying. An added advantage is that such a GIS is relatively simple to create and hence a lot cheaper than more complex ones (Ahmed Shaig, 2001).

According to Peng and Tsou (2003), a static map publishing refers to embedding maps as graphic images like GIF, JPEG, and Portable Network Graphics (PNG) inside an HTML page. The map images are used as a visual presentation to illustrate the points inside the HTML text. But the map image itself is not intelligent. That is, the map image is a static image displayed on the Web browser. The user cannot click on it to zoom to a certain area or get more information. A static map publishing does not support feature data at the client side and does not have map-rendering tools. It is a very thin client application that only supports ready-made map images on the Web browser. To publish a static map image, you can save a ready-made map as a graphic map image format and embed it inside an HTML page.

Client

In a communications network, the client computer is the requesting machine from the supplying machine, server. Peng and Tsou (2003) have been classified clients as two viz., Thin and Thick. Thin Client, where is little or no logic processing at the client side; most of the processing is performed at the server side. The client is merely used as a user interface for the user to input requests and to view processing output from the server. The client computer does not need to be very powerful, but the server needs a very powerful computer. Thick Client that most logic processing is performed at the client side. The powerful client - side computer is needed to handle client - side processing.

The client in this web mapping is the simple web browser with no client-side plug-ins or Java applets. The sole purpose of the web browser is to interpret the contents of the HTML documents that sent by the Web server and display them graphically. The web browser also helps navigate from one page to another using the embedded hypertext links. For the present study, the client/user are to be students and spatial decision makers like Estate maintenance and administration. As far as this project concern the user will be Think client. When the administration wish to develop the Distributed GIServices to Bharathidasan University then this work will be improved with effective GIS data server.

Create URL

To access a web page user/client need an unambiguous way of defining links to other documents or other Internet resources. This is done using a Universal Resource Locator, or URL. Peng and Tsou (2003) stated that URL contains information on the specific location of target resources. In typically consists of four parts: the protocol scheme, the domain name, the port number and the location of target resources. Protocol scheme specifies the type of protocol to be used to access the resources on a server. The domain name is the server name to identify the Web server site. A port number identifies the program that runs on a server. The location of target resources specifies the path and the name of resources: document, images, titles and so on. URL is an informal tern associated with popular schemes: http, ftp, mailto etc. Finally, the URL contains information the specific location of target resources. It is a hierarchical description of a file location on the computer. It includes a file directory, subdirectories and file names.

Static Map Publishing Through Blogger

Blogger was started by a tiny company in San Francisco called Pyra Labs in August of 1999. It started by three friends, funded by doing annoying contract web projects for big companies, trying to make their own grand entrance onto the Internet landscape. Google and Blogs were amenable to the idea. Now they are a small team in Google focusing on helping people have their own voice on the web and organizing the world's information from the personal perspective. It supports to determine the web server, either FTP or STP, besides that allows make separate URL through connect TCP IP/IP address of server, which sends data on basis of client request during access.

Blogger Browser Matrix

Blogger's awesome WYSIWYG (What You See Is What You Get) editing features are fully supported in the latest versions of the most popular web browsers: Internet Explorer 6/Windows and Mozilla Firefox. Safari support is coming, eventually. Blogger's Preview mode and HTML formatting buttons work in some older browsers, but not in others. Table shows what works where.

Key

Browser: web browser name and version.

Compose: whether or not Blogger's wysiwyg Compose mode is supported

HTML buttons: whether All, Some or None of Blogger's HTML mode formatting buttons are supported.

Preview: whether or not Blogger's Preview mode is supported

Usage Breakdown

- 71 per cent of our users use IE6/Windows; clearly the majority yet a lower percentage than the web in general.

- 6 per cent use Mozilla Firefox; an up-and-comer, which we all use internally.
- 3 per cent use Safari; also widely used internally.
- 2 per cent use AOL

Browser	Compose	HTML buttons	Preview
AOL 9/Mac	No	None*	No*
AOL 9/Win	Yes**	All	Yes
Camino 0.8/Mac	No	All	No
Firefox/Win	Yes	All	Yes
Firefox/Mac	Yes	All	Yes
IE 6/Win	Yes	All	Yes
IE 5.5/Win	Yes	All	Yes
IE 5/Win	No	None	No
IE 4/Win	No	None	No
IE 5/Mac	No	None	No
Mozilla 1.7/Win	Yes	All	Yes
Mozilla 1.7/Mac	Yes	All	Yes
Mozilla 1.3/Mac	No	All*	Yes*
Netscape 7.2/Win	Yes	All	Yes
Netscape 7.2/Mac	Yes	All	Yes
Netscape 6/Win	Unsupported		
Netscape 6/Mac	Unsupported		
Netscape 4/Win	No	None	No
Netscape 4/Mac	No	None	No
Opera 7.5/Win	No	Some***	Yes
Opera 7.5/Mac	No	Some***	Yes
Safari 1.2.1/Mac	No	Some***	Yes

AOL users, as well as those unaccounted for here (Netscape, Opera, etc.), would be wise to use Mozilla Firefox in order to have the best possible experience using Blogger.

Web Site Creation in Blogger Environment

In the present study an attempt has been made to frame distributed GIS for Bharathidasan University using Blogger.com, which is one among familiar internet platform www.DistributedGIS.blogger.com It is also known as the Push Button publishing, where separate site has been created for publishing the static map of the Bharathidasan University. While publishing the static map PDF format is suitable for downloading the maps rather than the viewing. Hence in this study all thematic layouts are published using JPEG and GIF format file formats. In addition to that the tables, photos, report has also enclosed.

For Query layout the criteria report and analysis also embedded with HTML document. The present study can apply for dynamic and interactive web mapping for Bharathidasan University in future that would be useful to planners to decision making and students to get more information relevant to their query.

Summary and Conclusions

GIS provides the ability to extract the different sets of information from a map and use as required. The development of GIS has been influenced by the progress of IT. The internet influenced GIS in three major areas, as data access, spatial information dissemination and GIS modeling/processing. Variety of Internet GIS is used for data sharing, information sharing and knowledge sharing (Ming-Hsiang Tsou, 2004). Distributed GIS represents a dramatic departure from the traditional two-tier client/server model and a broader framework including both internet GIS and mobile GIS.

The main aim of the present study is to publish static map of Bharathidasan University for students as well as administrators and disseminate distributed GIS. This work would improve university management as well as disseminate the Distributed GIS - static map publishing through user-friendly approach.

The study area is the Bharathidasan University North Campus at Palkalaiperur, which lies between 10° 39′ 44″ and 10° 41′ 12″ North latitudes and 78° 44′ 16″ and 78° 45′ 2″ longitudes and located in the Tiruchirappalli district. The University is a member of the Association of Indian Universities and Association of Commonwealth Universities. It is entitled to receive grants and donations from University Grants Commission (UGC) and other agencies all over the world.

GPS Survey has executed for entire University Campus (1000 acre) using Leica GS5. All point and Polyline features are captured and assigned attributes to them. For effective mapping purpose it had to integrated GIS and GPS for effective mapping. Then integrating all these information with respective high-resolution satellite image to pertain synoptic view of the university area. Finally, based on this information map are generated for students and administrators on the basis of spatial query using ARCGIS 9 Version with evolving a Distributed GIS for University. At last a separate URL has been created using Blogger.Com internet platform for publishing static maps to access the clients, who are thin clients. Because it supports several recent web servers with future coming servers.

In present study which problems to be solved are identified that includes Administrative related details, which are not spatially defined, with the facility of updating. So effective administrative infrastructure management is not ease in most of the situations. Hence it is necessary to manifest and manage by spatial decision supporting system. In addition to that spatial management dissemination is a right to public also. Not only for administrative purpose but also the concern on other user of students who wants to get familiar of basic infrastructure of university. Those users may be in abroad or other state. To solve this problem the Distributed GIServices has found to help in paramount manner. Distributed GIS - Static Map Publishing is the web mapping technique with two-tier (Server/Client) approach. So this is preliminary wok for Distributed GIServices to the Bharathidasan University. This work will be enables the spatial service for decision support and disseminate the information related academic to the planner as well as the users. In future it will develop as Distributed GIServices for Bharathidasan Univeristy.

REFERENCES

Ahmed Shaig, (2001). An Overview of Web based Geographic Information Systems, Presented at SIRC 2001 - The 13th Annual Colloquium of the Spatial Information Research Centre, University of Otago, Dunedin, New Zealand, December 2-5, 2001.

ArcGIS 9 S/W, Desktop help, ESRI, Product.

Chang Kang - Tsung (2003). Introduction to Geographic Information Systems, New Delhi: Tata McGraw-Hill publishing Company limited. 110008. (pp 13, 134 - 150, 160 - 168, 183)

Claus Rinner, (2003). Web-based Spatial Decision Support: Status and Research Directions, Journal of Geographic Information and Decision Analysis, 7, pp. 14-31.

http://www.business.otago.ac.nz/SIRC05/conferences/2001/36_shaig.pdf

http://www.geodec.org/Rinner.pdf

http://www.gmat.unsw.edu.au/snap/gps/gps-survey.html

Ian Heywood, Sarah Correlius, Steve Carver (2001). An Introduction to Geographical Information System. Delhi: Pearson Education.

Lo. C.D., Albert K.W. Yeng, (2002). Concepts and Techniques of Geography Infromation Sytems, New Delhi: Prentice - Hall of India Private limited.

Manjula, and S. Aruchamy, (2004). Resource Atlas of Bharathidasan University and its Environs, TamilNadu, India, Using GIS, Unpublished M.Sc., Dissertating Tiruchirappalli: Bharathidasan University.

Ming-Hsiang Tsou, (2004), Present Realities and the Future of Internet GIS, GIS Development Journal, 8, pp. 29 – 32.

www.blogger.com

Zhong – Ren Peng, and Ming- Hsiang Tsou, (2003). INTERNET GIS Distributed Geographic Information services for the Internet and Wireless Networks. New Jersey: John Wiley and Sons.

INDEX